Urban
Self-
Management

Simona Ganassi Agger

Urban Self-Management

PLANNING FOR A NEW SOCIETY

M.E. Sharpe INC.
WHITE PLAINS, NEW YORK

Published simultaneously as Vol. IX, no. 1-2 of
International Journal of Sociology.

Library of Congress Catalog Card Number: 78-73223
International Standard Book Number: 0-87332-125-1

Printed in the United States of America.

This work is an expanded and revised version of the Italian edition,
Autogestione urbana: L'Urbanistica per una nuova società, published
by Dedalo in 1977.

To Bob,
whose respect for my ideas and
whose help in clarifying them
contributed to my understanding of
what maieutic approaches can mean
for democracy.

TABLE
OF CONTENTS

ACKNOWLEDGMENTS

I would like to thank Ignazio Malocco, Manuela Zevi, and Flaviana Meda for their help in producing the Italian study on which this book is based. If this work now meets the needs of American and Canadian readers, a good deal of the credit should go to Paul Piccone. If it does not, the blame is mine for failing to carry through on his suggestions. Finally, I would like to thank Arnold Tovell for his contributions, which went above and beyond those normally to be expected from an editor.

Simona Ganassi Agger

Urban
Self-
Management

ALTERNATIVE URBAN PLANNING:
POSSIBILITIES AND CONSEQUENCES

This book has a relatively simple thesis: institutions must and can be opened to people so that they may begin to participate in matters of vital concern to them, matters that are now regarded as the concern of experts. One such set of experts, of professionals, is the group identified as urban planners. In fact, I am among them as a professor and a practitioner of urban planning. Our domain is the organization of cities and of urban (as well as nonurban) places and spaces. My thesis is that urban planning must and can be opened so that ordinary people practice urban planning. The beginning of a revolutionary transformation of society may start from such an opening, and I argue here that urban planning is a particularly appropriate domain in which to start building toward a new society.

We start with some results of an international study that bear on the problem of citizen participation in urban planning and which give us reason for optimism if the appropriate lessons can be learned from the study. Then we look carefully at what urban planning is supposed to be and what it is in fact. Before we are through we shall have also presented a picture of what happened in an Italian experiment in trying to obtain increased citizen participation, because crucial lessons were also learned from that. In particular a methodology was developed wherein urban planning was demystified, the crucial step toward a truly participatory planning process. In the course of that effort I began to appreciate how professional urban planners may begin to have an important new role in a new kind of urban planning process, a process I call "maieutic" planning. This role is one of helping others, nonprofessional people, to understand how they are already engaged in planning and how they may become more productively engaged. But the reader should remember that this is essentially a treatise on how to move toward a postindustrial society different from those currently projected rather than a treatise on urban planning or urbanism per se.

Participation: Real Questions and False Problems

The goal of opening institutions to the participation of ordinary citizens is, I believe, the nodal point for producing innovative transformations in modern society. Such a goal, however, evokes a host of objections, some of them needing to be taken seriously.

Some objections are made by those who are decisively opposed to broadened participation because it is against their interests. They are usually the ones holding power and are more concerned with this than with resolving or improving the problems of modern society. Bluntly put, these critics do not concern me and I shall not respond to them.

There are others, however, who maintain that society cannot be transformed through more participation. In order for there to be radical change, the economic and political power structure must pass into the hands of those representing other classes, those which are presently powerless. Such change can only be prepared by a counter-elite that constantly must control and guide the powerless. Thus, participation does not constitute the fundamental ingredient either before or after the supposedly radical transformations of society. The basic presumption here is that those who have these guidance roles will represent the interests of the masses. Thus, participation by everybody is not necessary and is, in any event most unlikely, until well after a revolutionary change.

On the other hand, there are those who say that they agree with the hypothesis of participation, but they define it in a different way. They consider it useful, just, and democratic that decisions be made after hearing the people's opinion or after informing them, but feel the actual decision should be up to those who, having special knowledge and roles, alone are in a position to make the final decisions.

And, finally, there is another criticism, made by some who agree in principle with the idea of participation, but who believe that today it is necessary to reject it as a matter of realism. Conditions for starting such processes simply do not exist. As past experience shows, the people themselves do not participate, are not interested and are not able to make decisions.

As I said, I will not try to refute the first group of objections. It is clearly useless and my objective is to present my ideas to those who have an interest in broadened participation or have an interest in accomplishing the social modifications demanded by the situation in which we live today.

Let us consider the criticisms of those who consider participation a second-order objective in the transformation of society. My argument is directly opposed: there cannot be a revolutionary process that moves very far based only on an elitist conscience.

On the contrary, only by finding and fighting for the forms that will begin to open immediately the decisional processes to all can one bring revolutionary innovations to society. Otherwise, there is the near certainty that the revolution serves only to change the identity of the power-holding elite and some elements of the social organization. Furthermore, even an elite that wins power in the name of the powerless has the problem of maintaining its power. That leads quite naturally to processes of repression that certainly do not respond to the requirements of renewal. An elite is, indeed, justified in maintaining elite positions if it believes that elite specialization is necessary for the general welfare and common good, but I reject that contention.

What about those who are in favor of participation but only in a "consultative" form? First of all, this is the traditional norm in representational systems termed democratic. As such, it is not an instrument of transformation; indeed, the extension of such consultation may result in more stabilization.

Besides, such a route appears truly impassable. One of two things may happen. The more likely is that people will become tired and cease being available for limited consultation. Or they will become available for more than mere advice-giving. If they choose the latter, it is quite natural for them to demand guarantees that their views not be ignored by those who "decide." This, it seems to me, amounts to transforming consultative participation into actual decisional participation of just the kind I am proposing.

I have reserved my reply to the most important criticism for last, precisely because of its importance. This objection is that of those who are in favor of participation but are skeptical and discour-

aged in light of the scarce manifestations of interest in participating
and the scarce results that have emerged from the experiences so far.
My rebuttal starts by recalling that experiences of participation so far
have rarely had the characteristics of decision-making involvement.
When there have been such experiences and when they have really
modified traditional decisional processes, they were brusquely cut
off. This has been true both for experiences in North America in regard to the "war on poverty" of the sixties and for the experiences in
Italy that will be discussed later.[1]

In any event, concern over the difficulty of getting people involved, especially people from the most marginal groups and social
categories, is not misplaced; in the modern world such participation
does not happen spontaneously. Although there are conditions push-
ing the so-called mass of humanity to enlarge its self-knowledge and
awareness of its own human status and rights, these conditions are
opposed by thrusts toward the fragmentation of social relations,
apathy, representation by proxy, and privatization.

On balance, then, there is substantial evidence of concrete problems to be addressed. It does not help, however, to reply by way of
renunciation or by the assertion that participation is a "utopian"
dream to be postponed to an unspecified future when the world will
be different. It does not help and it is not necessary.

If there are people who really see the need to change society,
but feel the moment is not now, they may, indeed, limit themselves
to interpreting and trying to satisfy the needs of the disinherited and
disadvantaged as best they can. But through their altruistic elitism
such people risk reinforcing the patterns of nonparticipation in the
society. One must begin today to forge the bases for desirable basic
changes. I am convinced, and I will try to show how, experiences in
decisional processes in the urban milieu can constitute a beginning of
a real participation that is already within reach.

It is precisely because this *can* happen that it is clear that one
must take into account the obstacles and conditions that today are
in the way of the active involvement of people, all of the people.

In examining the real problems of an alternative process, as I
propose to do in this chapter, one must clear the field of a widely
held assumption that in reality is a pseudoproblem: that is, that people do not have the capacity to participate, that they are not sufficiently prepared or are not competent to take part in decisional processes.

This assumption is widespread and for that very reason dangerous because, in effect, it thoroughly distorts reality. In urban planning, assertions of this type are used to support the following line of argument: by opening the decisional process, one risks the possibility that some skilled speculators, with very precise interests, may succeed in manipulating others to the point of convincing them that what is good for the speculators is in the general interest. The equation becomes citizen participation equals citizen manipulability equals selfish speculative development.

It is also asserted that what usually happens is that ordinary people, lacking other models than those dominant in the present society, use criteria that in the end can only reproduce the current consumer society. In other words, their choices will be closer to the speculative interests of the few than to the nonspeculative interests of the many. Greater citizen participation at best would therefore lead to the same kind of urban society or even a worse one than exists today.

Let us examine these arguments. The fact is that technicians and administrators have affected the organization of physical space, especially in the thirty or so years from the end of World War II to today. As I will show in Chapter Three, the distressing models of consumer society, even at the level of the organization of physical space, have been reinforced by the images and physical organization produced by these traditional procedures.[2] Thus, if we want to be consistent, it is the incompetent technicians and administrators, acting either purposefully or unwittingly for vested interests, and not the people, who should be excused or removed from the task of decision-making in the future.

One may object, though, that given an ever more complex society, it is impossible to allow the common people, who have no specialized knowledge, to decide in such important matters as urban planning which in turn influence the local economy, finance, services, and so forth. Without adequate methodological tools and specialized knowledge, how can a common citizen, obligated to make a decision, understand the effects it may have on a distant future?

Such a concern is understandable. It follows, though, that it would be important for every citizen to understand the meaning and consequences of what is happening. This ought to be posed as the crucial objective and, as a result, the necessary means to attain it should be sought.

Furthermore, why don't these concerns also apply to the other sectors of public life? For example, if we put together all the people involved in public, governmental decision-making from the municipal to the national level in any country, especially from the end of World War II to today, we would discover that the overwhelming majority are "nonspecialists," persons who learned only through experience how to be aldermen, mayors, congressmen, secretaries of departments. Someone may respond, however, that although these people learned "on the job," they had a large quantity of specially provided and often predigested, condensed and simplified information as well as the help of "experts," bureaucrats from municipal and state or federal administrations who furnished them with knowledge and experience. Precisely! I do not see why this cannot also be done for ordinary people. To begin with, they should be supplied with that information and knowledge that usually does not reach them because such data are classified as "secret" or simply because the trouble is not taken to give it to them. This information constitutes a part of the decisional process itself. The lack of it becomes one of the elements that in so-called citizen-participation meetings puts the people in the room at a very different level from those who are behind the table with their glasses of water in front of them. As Ralph Nader has said, even university graduates know almost nothing and have been taught almost nothing about how even their local governments work.

The fact that the people do not respond "adequately" is, thus, determined by external conditions and not by the real potentialities of people. They have not been in a condition to be actively present in the decision-making process except as people to be manipulated by others. Indeed, the problem of "adequacy" is a pseudoproblem. Certainly the thesis that people generally are not sufficiently capable or competent to participate cannot be demonstrated by the results of past experiences with citizen-participation efforts.

What has happened so far in those efforts is important, nevertheless, because it shows the insufficiency of a strategy centered only or predominantly on invitations to participate in public meetings.

The basic questions, the real questions, suggested by the present situation do not concern the capacity of people to make a valid contribution. Rather they concern the present lack of involvement in civic and public affairs and the seemingly slight interest in general problems. In fact, it frequently appears that what at first seems to be a general interest is rapidly reduced to a very narrow interest in matters or aspects connected to immediate and personal questions.

But the problems arising from the lack of interest are also usually posed in a superficial way by even those who advocate participation, who often limit themselves to maintaining that participation is slight everywhere, even in so-called socialist countries—at least when it is not compulsory. Few stop to consider or analyze who participates—that is, what are the personal and social characteristics of the more socially involved persons? And among those with social characteristics that seem to make for low participation in general, who participates—who are the exceptions? It is difficult to discover the conditions of the social system and the actions that seem to have a major influence on the presence or absence of the people. In other words, the data, however "hard," on the low degrees of involvement in these spheres are not sufficient. Such aggregate data and such general facts are of little help in formulating a strategy for putting people in a condition to participate.

It is true that contemporary, especially capitalist, society rewards individualistic interests, and interests that divert attention away from potentially controversial civic, public and political involvement. It is equally true that in no country, not even in the most advanced industrial, capitalist societies, do we yet find ourselves faced with masses of people so apathetic or hallucinated as to be beyond all possibility of being induced to participate. The tens of thousands of persons who attend sporting events and take part in collective forms of so-called amusement show that the potential exists but has been routed into other forms.

This is not due to chance. Nor is it evidence that human nature itself is responsible for the growth of sports and recreation of a spectator variety primarily and for the decay or stunted growth of participation in civic affairs. On the one side, there are forces involved in using refined tools and techniques to transform man into a more perfect consumer of sporting events as well as of clothing, cars and detergents. On the other side, there are the so-called alternative forces, proposing a new society, but giving more attention to the mechanism and problems of power than to the plight of people. At the most, the "alternative" forces are dedicated to producing "good administrators" for those communities or countries in which they come to power rather than being deeply concerned with trying to involve people generally in becoming good "self-administrators."[3]

The results that this way of "good government" produces are not only insufficient for meeting the problems of modern society, for opening and going toward a society different for its quality of social life and for its equality; they are actually counterproductive as well.

In order to put aside this pseudoalternative of a people-oriented, beneficent but still elitist path once and for all, we must concentrate our attention on the problems of participation and establish where it is really opportune for one to begin in order to get results. The relevant data are scarce.

For this reason I will anchor my picture of the reality of contemporary participatory experiences on a study that, if not unique, is at least exceptional—a comparative international study involving eight communities from six nations representing both socialist and capitalist systems. From the former system there are communities in Czechoslovakia, Yugoslavia and Poland; from the latter, communities in Canada, Italy and the United States.

There are several reasons why this study is worthy of particular interest in the present context. One is that we are able not only to focus on the personal and social characteristics of those who do and those who do not participate in various ways and domains in these communities, but also to obtain a deeper understanding of such social systems in relation to participatory activities and the forms such activities assume in them.

As I will illustrate later, the presence of two Yugoslav communities in the comparative study is particularly significant in this regard. The study has also helped us not only to understand dynamics that differentiate the North American and Italian situations from that of socialist countries, but also to begin to understand how different at least two socialist countries are in regard to quantity and quality of citizen participation.

An equally important reason for our interest in the study is that it was conceived as a rigorously comparative investigation of civic involvement at the level of the local community. From the outset this led to particular care in assuring the uniformity of methods and the comparability of data to the greatest extent possible.

There is still another important aspect of this international community study that is of interest. The study differs markedly from available knowledge often drawn from work aimed at a broad knowledge of a particular community, touching on a vast range of elements and for "scientific" purposes, that is, knowledge for knowledge's sake. This study was, instead, directed to the problematic of citizen participation and aimed to be a support for actions designed to increase such participation in civic affairs. This aspect of the study is directly connected to another of my aims: that of examining in this book the possibilities for immediately initiating participatory actions rather than postponing them to a hypothetical time in the future.

Such an action began at the end of 1974, when a plan for re-vitalizing the historical center of Faenza was commissioned. About one-fourth of the total population of this city of 54,000 in the region of Emilia Romagna in northern Italy live in that historical center.[4] The planning process was intended to produce a truly participatory and self-managing plan from its very beginning. It started with a social survey that sought to gather elements of knowledge comparable to that of the international study, but the Faenza research phase had the distinctive character of being action-research aimed at constituting not only a piece of scientific knowledge but also at being an important first action of self-knowledge on the part of the citizens of the community itself. In Chapter Five I describe the evolution of this experience and examine its crucial aspects.

Elements for a Strategy of Participation

To begin our investigation, a brief survey of the international study will be useful. This study involved teams of social scientists from the six countries working together for nearly a decade. It took nearly three years of preparation, pilot-studies and instrument development before teams from the first four countries, Canada, the United States, Czechoslovakia and Yugoslavia, entered the major phase of doing random-sample surveys of one community from each country (except in Yugoslavia, where both a community in the republic of Slovenia and one in the republic of Bosnia were surveyed). Citizens' involvement in a multitude of domains in their communities was the central topic of concern, with particular focus on adult citizen involvement in education, an important and relatively noncontroversial subject for such comparative research then. In the early seventies the Polish community and two Italian communities were added to supplement the emerging picture of citizen participation in socialist and capitalist countries. In fact, the international team hoped to proceed to a phase of trying to induce more citizen activity in their community affairs but that phase was accomplished only in the Yugoslav communities and, more recently, in an additional, third Italian community.

The investigation concerned the following eight communities:

Bowmanville, in the province of Ontario, Canada

Guastalla, in the region of Emilia Romagna, Italy
Scorzè, in the region of Veneto, Italy
St. Helens, in the state of Oregon, the United States
Horice, in Czechoslovakia
Trzic, in the republic of Slovenia, Yugoslavia
Konjic, in the republic of Bosnia, Yugoslavia
Lubraniec, in Poland.

These communities are roughly similar in size. They are all small, with about 5,000 to 15,000 inhabitants. The two Yugoslav communities belong to regions that are very different from each other in terms of their historical past; this is also true of the two Italian communities. In each case they constitute sociocultural environments that also are commonly held to be historically very different. This has constituted a further basis of comparison for producing a more complete analysis of these two pairs of communities in their national situations.

The most important of these introductory comments, however, concerns the fact that in presenting the results that have emerged for each community, I do not want to assert that they constitute the prototype for the nation to which they pertain. Nor do I want to minimize the fact that it is impossible to identify even all of the major variations existing within each nation by relying on a sampling of one or two communities.

Nevertheless the data that have emerged illuminate or point with some validity to certain patterns of sociopolitical involvement that are specific to these countries and to these systems.[5] Those that I report were discussed with scholars, and also accord or are not inconsistent with the best social science literature, from each of those countries, and I therefore often use the names of the nations rather than those of the communities in what follows.

Similarities and differences not only between the two systems but also within each system or within a single country (as in the case of Italy and Yugoslavia) will be presented so as to help the reader independently to establish and evaluate their real significance with respect to the chances for direct action in the country in which the reader lives and acts.

For the sake of clarity in this regard, the exposition is organized according to three principal points. First the existing structures of participation are reviewed. Then the elements that in the current

situation appear to impede participation or characterize nonparticipation are discussed. Finally the elements that seem to indicate the
possibility of, or constitute incentives for, the presence of citizens in
the decisional processes of the community are discussed.

Structures of Current Participation

Obviously, the initial hypotheses of the international study
needed to be evaluated in terms of the actual findings. It must be
said, however, that we began with specific ideas about the characteristics of the communities under examination. It was first thought
that we would find ourselves faced with two different models; one
would represent the socialist countries and the other Canada, the
United States and Italy, that is, so-called capitalist countries of the
West. Due to historical-cultural reasons, one could also expect differences to emerge between Canada and the United States on the one
hand and Italy on the other but within a single underlying model.

Instead of that picture, the results of our study suggest substantially different interpretations. First of all, it was not possible to
interpret the three socialist countries by means of a single model. In
the second place, we could not interpret the situation of the Italian
communities of the study as a simple variation of the model of the
two North American communities despite the common capitalist system in the three countries. The result was the emergence of four
models for the six countries of the two systems.

It is worth emphasizing here that we are not in a position to say
if what has emerged from our study of the two Italian communities
allows us to specify a model valid for all of Western Europe or just
for the Mediterranean countries of Europe, or if it allows us to gen-
eralize even about the one country of Italy. With respect to the
Italian situation, we must raise a further question. Given the very
great and important differences between the north and south in
Italy, the picture that has emerged from our study of two northern
communities is better thought of as perhaps representative of northern Italy, but it would be too risky to go further, even tentatively. It
is our essentially intuitive feeling that such a difference is greater
than regional differences in the other countries (including the English
and French parts of Canada).

Our study, in other words, has pointed out in a more precise
way, one in contrast with the original hypothesis, that the label "a

country of the capitalist world" is not sufficient in describing the comparative situation of participation in each of our three capitalist nations. Perhaps more surprising, it is not enough to speak of "socialist countries" to give a valid picture in this regard of the nations of that system, even within the single area of Eastern Europe.

In other words, important social, cultural and historical factors differentiate the structures of participation, qualifying the more general factors that shape these major politicoeconomic systems.

Since a general report on the international investigation is not my primary purpose in this chapter and in this book, I will not discuss here the methods used to collect and to elaborate the data nor the detailed relations between and among the many variables. For that I refer the reader to the documents that have already been prepared and to publications in preparation.[6] I will merely mention findings that are relevant to our need to understand present structures and future possibilities of participation.

Interpretative models of the class structure of civic involvement in the communities of the various nations. The comparative international study was based on sample survey research in which two to three hundred or so randomly selected citizens over eighteen were interviewed in each community. The principal and basic data, including the primary data on participation, came from their answers to these relatively long, detailed and comparable interviews. There was every reason to believe people responded truthfully to questions about participation.

One set of important participation data concerned how many of these adult citizens were involved and how much, if at all, in eight major areas of community life. A second set of data concerned the kinds of formal associations in the community to which the people in these random samples belonged, that is, political parties, labor unions, churches, cultural and recreational associations, and the like.

Those data were examined with respect to the class status of those who did and those who did not participate. In other words, the characteristics traditionally considered to be the indicators of a person's social-class position and background were taken into consideration both for those present and those not present in community activities: level of education, occupation, and income, and the occupation and educational level of their fathers. In addition, a range of other data were examined, simultaneously, such as interest without more active participation in the several kinds of community affairs.

The four models for the six countries, mentioned earlier, are distinguished from each other according to the variation of one of two or both of the following coordinates: the degree of participation and the class characteristics of those who participate. Let us examine them in order, starting with the two communities in the United States and Canada that do fit a single model.

The North American Model. Contrary to what various other studies and popular impressions have held, participation in the North American situation is slight from the quantitative viewpoint and highly structured along the lines of social class.[7] In other words, participation in community social activities is largest among those belonging to the upper or upper-middle social class, both in Canada and in the United States.[8]

The power-holding decisional elite, made up of those with the most education, the most important occupational positions and the largest incomes, is divided into various sectors in a rather specialized manner. In other words, few participate in all sectors, but because of strong class bonds extant among the members of this group and their substantially common ideology, their presence, although sectoralized, constitutes a homogeneous hegemony over the life of the community.

Later in this chapter we will try to evaluate the North American situation more closely and draw from it hypotheses concerning possible strategies for effecting initiatives of citizen participation there.

The Traditional Socialist Model. There seemed to be many points in common between the Polish and Czechoslavak communities; their situation reflects what some studies describe to be generally typical of socialist countries of Eastern Europe today.[9] In these communities, too, the level of participation is not high. Indeed, it is very low in civic affairs proper—local governmental-political-economic-urban affairs—but participation in the community's cultural affairs, unlike the North American communities', is extraordinarily high.

It is clear that class, as we know it in capitalist countries, is overcome in these two socialist communities. Participation is connected more to personal experience, to interest in politics and to party involvement or to personal educational achievement rather than being conditioned by factors such as occupational status, income and social background. Personal interest in local civic as in local political affairs does not depend on one's social class and is not necessarily shaped by either social status or wealth. In fact, in the Czechoslovak community we found a negative relationship between the father's level of education and the interviewee's degree of participation in social life. Nevertheless, we must emphasize that what we have called the traditional socialist model describes a situation in which participation is slight and the structures of society, even on the local level, seem to be generally distant from the people.

The Yugoslav Socialist Model. Neither of the two Yugoslavian communities seemed to fit the terms of the previous model. This was especially true in relation to the proportions of citizens participating, which, in all except for cultural community affairs, proved to be much greater than in the preceding cases. The two Yugoslav communities do share with the Czechoslovak and Polish ones the characteristic of not having a traditional class stratification shaping participation. Education is important, again, but it does not relate to social background as it does in the West. The percentage of participants, though, makes it possible to conclude that the Yugoslav situation undoubtedly is much more positive than in the other socialist communities. That is, we find a model where the public institutions are closer to the people and more open or, vice versa, one can say that the people are more involved and present in all domains of community activities except that of culture.

We will have the chance to see other aspects that characterize this situation in the two Yugoslav communities—for example, the sense of being able to influence community affairs—which most certainly constitute important pointers for stimulating participation.

With respect to the situation of the two Yugoslav communities, though, we must note again that these areas are very different in their historical, social and cultural aspects as well as in their present economic situations. The Slovenian community, Trzic, is stable and economically very well off. It has a relatively long past of industrial development. The Bosnian community, Konjic, is basically a much poorer community, in rapid development, in the process of modernization and industrialization. The Slovenian community is predominantly of the Catholic tradition whereas the Bosnian community is characterized by the presence of several religions, in which the dominant and important one, however, has been Muslim. And while Roman Catholicism is still pervasive in Trzic, atheists outnumber others in the younger and politically more enthusiastic Konjic. Our hypothesis, then, is that a model of participation fitting both of these communities is probably valid for all of Yugoslavia despite its regional, ethnic and even linguistic diversities.

The model is closely connected to or explicable in terms of the most significant and interesting aspect of the modern Yugoslav experience: that of self-management, primarily in the productive sphere. The emphasis on *economic* self-management can be considered one of the factors that has limited and limits the still very interesting Yugoslav situation. But it appears clear from our comparative findings that the stimulus to take part in decision-making that people absorb in their experience at work does not remain limited to work-related decisions even though the people do not have the opportunity to the same extent to manage their own affairs in the sectors of civic life, despite official rhetoric about community self-management.

The Italian Model. We have said earlier that the model suggested by the two Italian communities is different from that of the other two communities of the capitalist system, one in Canada and the other in the United States. In this case, however, the difference is not a marked difference in the degree of participation as it was in the socialist communities. The low number of those participating in all three capitalist communities is similar although the participation of people in various community activities tends to be somewhat lower in the Italian communities than in the North American ones, even with respect to political activity.

The fact that political participation is low might be a surprise for many, since there is a widespread belief that Italians are highly politicized. Indeed, the Italian situation is marked by the active presence of traditionally competitive political parties, not only in those sectors where party division more traditionally exist but also in the universities and in various cultural affairs.[10] But our findings are really not surprising: other studies testify to the scarcity of participation in Italian politics so that in this important regard the two communities in the study constitute not atypical cases for an Italian model.

Guastalla is typically run by a Socialist-Communist administration and at the end of the last century even had one of the first Socialist mayors, the first among small cities, whereas Scorzè, the other community, has always had a Catholic and Christian Democrat tradition and is now governed by a very large Christian Democrat majority.[11] Yet in both these politically different towns there is as little participation in party activities as in the other civic sectors. This participation, then, is independent of the historical political tradition and the forces currently governing community life. We will return in Chapter Five to what this has meant for the kinds of citizen participation actions undertaken in Italy.

What does distinguish the Italian situation from the North American one is the fact that in Italy a strong class structure linked with differential participation does not now exist. The class structure itself still exists but its earlier link to community participation has been broken. This is a striking contrast to the American and Canadian patterns.

Along with the differences between Guastalla and Scorzè a common characteristic emerged.[12] The highest social class according to indices of occupation and income does not figure prominently among those who are more actively involved in the key political, economic and urban sectors of community life. In most instances one finds those with modest occupations and average incomes among the persons most involved; the occupational categories represented are blue- as well as white-collar work. And this does seem to be a difference in class-participation dynamics rather than in class per se, since, for example, education is associated both with occupation and income in Italy as it is in the United States and Canada.

This difference in regard to class is surprising to those who imagine Canada and the United States as classless or weakly stratified societies and Italy as a highly stratified one. The fact is, however, that in the two Italian communities, especially in Guastalla, the relationship between present social position and social background is not as strong as in the North American communities.

The important fact of scarce participation does not constitute evidence of the traditional capitalist situation still existing in Italian communities; poverty of participation also characterizes one of the two models of the socialist community. But in the Italian communities the lowest classes are not as predominant in community activities as they are in the Polish and Czech communities.

The traditionally dominant classes, the upper and upper-middle classes, no longer have exclusive control at the local level in the Italian as they do in the North American communities. In the Italian model there are among those most involved in civic affairs classes until now excluded in the North American model. Nevertheless, this does not allow us to conclude that the Italian communities are either more open or egalitarian than others even though there is a more nearly equal presence of all social classes. Various middle and lower groups and classes as well as the most marginalized no longer are "represented" by the traditional power elite but still remain in a condition of being represented rather than being present themselves. Here, then, the ongoing question is that of understanding how and in what way it is possible to provide everyone the right of *self*-representation.

The scarcity of participation in six of the eight communities suggested a deeper examination be made of those who do not participate in any form in civic affairs and of those who are estranged from social life. This absence, moreover, includes absence also from membership, and especially active membership, in the community's formally organized associations.

In particular, the brief analysis of those who do participate implied the social class (in communities of the capitalist world) and social characteristics (in communities of the socialist world) of those who do not participate were the inverse. For this reason further examination was made of such sociodemographic characteristics as sex and age to see if they could be of importance in singling out predominantly or totally marginalized groups and, if so, to see if there were also particular models of behavior relative to those groups for the communities of the two major kinds of systems or for their respective subsystems.

Before entering into that matter, some further general information is necessary. In all of the communities studied, a high correlation was found between the two types of participation on which we were gathering data: participation and involvement in community actions and participation in the so-called formal associations of the community itself. In fact, they correlated in all samples.

In the following pages, therefore, I intend to draw from the international study information that is crucial for elaborating participatory strategies. I will primarily focus attention on involvement in local governmental and political activities, urban affairs and community economic policy rather than in other civic domains.

Marginalization in its Sociodemographic Connotations. The fact of scarce participation acquires further significance when one examines the patterns according to the sex of the persons interviewed. In general, it was found that in all areas of community involvement and in all the formal associations (excluding religious ones in some communities) women were less often present than men. The magnitude of such differences varied from community to community and from domain to domain. Active participation by women in matters directly connected to local government and politics was also generally far less than it was for men. The situation does not change if we examine only men and women belonging to the labor force rather than all men and women.

In socialist countries one can make more sense of that situation in light of the fact that involvement in these areas is highly correlated with political interest and with the fact of party membership. In three of the four socialist communities (the Polish one and the two Yugoslav ones), women show an interest in politics, but their presence within the party is proportionally much less than that of men. In the fourth, the Czechoslovak community, scant membership by women in the party goes with very little political interest but not any less interest than men evidence in political affairs. Socialism per se has not yet transformed the masculine world of local politics and government into a sexually egalitarian world. Only in Czechoslovakia is the interest in that world equal, but apparently only because that world has very little interest even for men. In Czechoslovakia even more than in Poland the local community's political and governmental structures are dominated from on high (via the party and the so-called National Committees that administer localities).

With respect to participatory actions, Konjic, the community in Bosnia, shows the least female marginalization of all four socialist communities. This testifies to the radical change that has occurred in this community, where women were historically subservient in the official doctrines of their religious culture whereas in other communities female subordination was more informal. One can say that the current interest of women as well as men resonates from the very rhythms of the transformation of the community. The economic, housing and general urban situations of Konjic are rapidly being improved, even though they are still far inferior to those of Trzic, the Slovenian community.

In the nonsocialist countries the common element shared by the two North American and the two Italian communities is, once again, the near absence of popular participation in governmental activities and local politics, especially on the part of women. Nevertheless, women in the Canadian and the United States communities are far more often present in cultural and recreational activities; and in the two Italian communities there is a greater presence of women than men in church-related activities.

One can, then, speak about the political marginalization of the women in all of the capitalist communities. It does not even vary with occupation; that is, it is not characteristic only of housewives.

We will return later to the interest, even if not actualized, in civic and political affairs expressed by women in these communities as well as to their interest in being more informed about events in which they are interested. In the meantime, it is important that we emphasize another point to which we will also return later. The women included in the samples interviewed in all eight communities, both Eastern and Western, generally have a lower level of education than the men. And, as we have seen earlier, in both the socialist and capitalist countries, the level of education is highly correlated to civic involvement. We will see that at the center of such a correlation there is a factor, which we have called self-confidence, in relation to which women generally are also lower than men. Thus, we can conclude that women everywhere still constitute a segregated and subordinated part of the community. There are no major variations in that regard from one political system to the other: it is the case within the context of the overall paucity of participation, and in the more participatory Yugoslav communities.

There is, however, a positive element here. With but one exception, a substantial number of women everywhere are interested in various aspects of social life, including political events that are traditionally dominated by men. (The exception is in Czechoslovakia, where scarce interest also characterizes men.) This allows one to understand that the situation can change: women, because of their already existing interests, can become a reference point for future projects to open decisional processes. One could not easily or justifiably come to this conclusion only in light of the data on the low levels of active participation of women.

The general exclusion of women from civic affairs is not the only such exclusion. Let us look at another one that is clearly visible when the relationship between age and participation is examined.

People within the samples were categorized according to three age ranges: young adults, those from eighteen to thirty; the middle-aged, those from thirty-one to sixty; the elderly, those sixty-one years of age or older. There was a significant presence in civic activities and in formal organizations for the first two age groups but it diminished rapidly for the elderly. Once again, it is impossible to identify uniform patterns according to type of political system. Let us examine community political involvement as an illustration.

The situation is variable for the young. They participate in local politics in the Canadian, the Polish, and in one of the Italian communities (Guastalla) to about the same degree as the middle-aged. In the United States as well as the Czechoslovak and Bosnian communities, they are less present. In Slovenia and in the other Italian community young adults are present to a greater degree than those of middle age.

As for the elderly, they are rarely politically involved. They are least present in the two Italian communities, followed by the Bosnian community. In the two North American communities, the elderly are a little less excluded (in the Canadian community still a little less than in the United States community). The other communities constitute the intermediate position between the most negative point, Italy, and the most positive, North America.

The data that have emerged on the relatively low interest in, and desire for, having more information about various aspects of community affairs support the finding that the elderly are least involved of all age groups. Once again though, one finds that in the two North American communities the elderly are more potentially involved than are the elderly elsewhere.

The matter of self-confidence brings into further relief the fact that the elderly are the most disadvantaged age group; that is, in all of the communities they are the least self-confident. It is important to note in this connection that in our samples the elderly in all of the communities had the least education; education breeds self-confidence in every one of our countries and communities.

From the picture that has emerged, the major difficulty in modifying the marginalization of the elderly appears clear. The similarity in their situations in both major types of political systems (indeed, the capitalistic North American and Italian situations, although not distant, are the two opposite poles) makes one important point. The expulsion from community life of those in this highest age group is not simply the result of the type of exploitation practiced on people in a capitalist environment.

In other words, we find that the disregard shown for the elderly and for their experiences or the nonconsideration of them because they are no longer a productive force and therefore no longer "useful" to society, finds expression in the socialist as well as the capitalist systems. We can say that it becomes a characteristic, even a philosophy, that penetrates all political systems because it permeates modern civilization, conditioning many of its values—regardless of the system.

Someone might argue, however, that the absence of elderly people from the community sphere of social life constitutes a natural phenomenon. It is due to physiological or biological facts of aging and it is precisely for this reason that the civic absence of the elderly does not vary by political system.

To test such assertions, it is useful to examine the results of an analysis of the Italian communities which took into account simultaneously the age of people, their level of education, their income and their involvement in processes of participation.

The results were that nonparticipation is manifested especially by the elderly who have a low level of education and income, meaning, of course, the great majority of them. Only a very small proportion of the few elderly who do have a high degree of education and income is nonparticipant.[13] This seems to be sufficient to put to rest the "natural" consequence of aging argument. It would be difficult to sustain the argument that the attempt to stimulate or afford opportunities for the elderly to participate in the life of the community should not be an objective.

The Factors Inhibiting Participation

So far we have examined the picture of the cross-national contemporary situation in relation to participation and have arrived at two different and important findings. First, in regard to participation and the socioeconomic variables, the communities of the international study evidence significantly diversified situations so as to give rise to several interpretative models although they are different from those of the traditional dichotomy of capitalist and socialist political systems. The second finding is that in regard to participation and the demographic variables, the similarities prevail over the differences, similarities that find in all eight communities women and the elderly constituting two social groups that tend to be marginalized. Further, these two groups differ from each other in such marginalization. In the case of women, almost everywhere there is interest in participation that, as we have suggested, allows one to presume that it will be much easier for their situation to be changed. In the case of the elderly, however, there is also a self-exclusion that results in disinterest.

With the exception of the two Yugoslav communities, then, a general scarcity of participation prevailed. Beyond that finding, which is not likely to surprise the reader, it is important to try to see what elements were inhibiting that participation in these communities.

The conditions of two things helped us significantly in understanding the present participatory situation of the communities studied. We will take them together because they constitute complementary data/factors: the optimism or pessimism of the persons in the samples with respect to being listened to by local government representatives, if those people went with problems requiring action by the latter, and the feeling of power or powerlessness in being able to exercise influence in solving community problems.[14]

First of all, let us examine these two attitudes in the eight communities. As for the first—whether or not the local government representative listens responsively—if we call the negative response cynicism, we observe that the least cynical samples are in the two Yugoslav and the Canadian communities, followed by that of the United States. Those in the Polish and Czechoslovak samples evidenced a substantially equal and high level of cynicism, while the most cynical were the two Italian samples.

A suspicious attitude toward public power is often considered an immutable constant in the Italian population, one derived from centuries of negative, exploitative historical experiences. As we shall see in Chapter Five, in the light of the results that have emerged in the other, separate Italian community study in Faenza, such an assumption is doubtful. Events, especially those closely touching the community, can influence this attitude. It is far from being immutable, even if it is now widespread.

In relation to the feeling that one is capable of influencing community decisions and thereby helping to solve its problems, once again the two Yugoslav communities reveal the most positive attitude. Then come the two North American communities. In this instance, we find the Polish and Czechoslovak communities at the lowest level.

The two Italian communities vary. Scorzè is rather close to the relatively positive levels of the North American communities while Guastalla is close to the relatively negative levels of Poland and Czechoslovakia.

By considering these two kinds of data together, we see that one pattern was found that fits the remarkably high participation level in Yugoslavia. There the latter is linked to a low level of citizen cynicism and to the highest level of citizens' feeling capable of influencing decisions. It is important to make a further observation in this regard. Except for one category (women with the lowest level of education in the Bosnian community), the general result of the feeling of local community decisional potency is incredibly confirmed by detailed multivariate analyses. Taking into account every level of education, both of the sexes and of the several age groups, the resulting subgroups in the two Yugoslav communities (with that single exception) are at a level of felt potency superior to all the other counterpart groups from all the other communities.

The feeling that one is capable of having an influence in community decisions and that the possibilities to do so exist proves to be closely associated with the level of education in nearly all of the communities, including the two Yugoslav communities. The Polish and especially the Czechoslovak communities constitute exceptions in this regard; there the highly educated feel just about as impotent as do the poorly educated.

When we compare those with high or medium levels of education in the four communities of the capitalist system, we find very similar levels of felt potency. In the two Italian communities, however, the number of those with positive feelings of potency is lower than in the two North American communities. That is due to two things. First, the Italian populations have lower levels of education than do the American and Canadian populations. Second, in both Italian community samples, including Guastalla where there was (and is) a left local government of Communists and Socialists, the poorly educated felt themselves to be without influence more than did their poorly educated counterparts in the two North American communities!

Let us now turn to the connections between these factors and variables. In all of the communities, except for the one in the United States, women less often felt they were in a position to have influence than did men. Women were also more pessimistic about local government representatives' listening to them.

With regard to the elderly, in all of the communities they were somewhat more pessimistic than those from the other age groups. The major feature of the attitude of the elderly concerning the possibility of being listened to by their local government representatives was the generally high percentage of those who said that they had no idea: they did not know or could not say. In other words, the elderly had a feeling of distance from their local government rather than cynicism as such. The proportions feeling capable of exerting some influence proved to be, compared with the less elderly, very low.

The consideration of these findings, then, leads to the inference that participation is actually influenced by these kinds of perspectives. Indeed, it is hard to believe that those who personally feel that their demands and opinions will not be taken into consideration can be induced or pushed to engage in civic affairs.

An Evaluation of the Level of Information. The data in the international study were collected on the basis of working hypotheses formed on the basis of the results of other studies, of literature in the social sciences, and on common sense.

One such hypothesis concerned the possibility of finding a connection between the degree of information that people have concerning the problems and events of the community and their participation in civic affairs. In fact, it is often held that such information is the essential and crucial basis for participation. By supplying people with more information on civic affairs, one begins to stimulate more participation. So goes the democratic faith.

In this study, therefore, two kinds of questions related to this hypothesis were posed: how many persons considered themselves adequately informed about community events, and how many were interested in receiving more information in each of the principal sectors of community life (asked about separately).

With regard to feeling well or poorly informed, the situations of people varied from community to community as well as within communities. Those variations, however, were not connected to variations in the degrees of participation. The two Italian communities offer a typical example of this. Similar both in terms of amount of participation and in the class structure of participation, they showed polar-opposite results. In the community of Guastalla there was the highest proportion whereas in Scorzè there was the lowest proportion of people of any community who declared themselves to be well informed.

Here we must underline the fact that the negligible correlation between the sense of feeling well informed and participation level is not as important for itself as for what it can mean for future efforts to stimulate participation. In fact, as everyone realizes but often forgets, being informed about an event is not equivalent to taking part in it, yet the mystique surrounding information appears to endorse the hypothesis that it is enough to supply relevant information in order to bring about widespread citizen participation.

It is true that knowledge about what happens in the community, and this includes matters hidden by official "bureaucratic secrets," is important. Still, our findings demonstrate that it is not sufficient to supply people with information for participation to occur.

Feelings of Inadequacy and of Incompetence to Take Part in Decisions. We have examined two responses that constituted indicators of decisional potency, one being the pessimism or cynicism of people facing their local administrators and the other, the feeling of personal powerlessness in influencing solutions to community problems. We considered such factors complementary, since both help us to understand the personal attitudes signifying the positive or negative relations between a person and the community to which he belongs.

The factors that we are going to examine now indicate, even more than the preceding ones, qualities of personal character, that is, what can be termed the psychosocial.

In this instance we are not dealing with data from only one response of the persons interviewed, but rather with a more complex analysis of a series of responses. In all of the communities those interviewed were confronted with more than one hundred statements concerning personal convictions or perspectives, such as "People are basically good" or "I would rather do familiar tasks than always to face new problems." They were then asked whether and how much they either agreed or disagreed to each statement.

Through factor analysis the responses in every community (and also taken all together) to twenty-four of the statements showed a strong intercorrelation pertaining to a single underlying factor (or subfactor) or dimension.[15]

Some examples of the twenty-four statements are "I feel uncomfortable talking in a group of people," "I would rather do familiar tasks than always to face new problems" and "There is very little I can do to change what life has in store for me." The factor or dimension manifested by these statements seemed to us to be one of *self-confidence* or *vitality*, appearing to have five facets: a capacity to act socially, a feeling of nonfatalism, a readiness to meet obstacles and difficulties, a feeling of nonconformity and a sense of tolerance.

The examination of this factor was of fundamental assistance in understanding something more about the behavior of people in our eight communities. It helped also to provide some clues for the further development of civic participation.

In the two North American communities this factor proved to be closely connected to the variables of social class and social background. The bulk of those who had high social status and also came from families with similar status had very high levels of self-confidence. The opposite was true for North Americans of lower-class status and background. By contrast, in all of the communities from the socialist countries self-confidence was dependent on more personal characteristics. In the two Italian communities the situation proved to be intermediate.

In all of the communities, however, the level of education was linked to this factor. In the case of the socialist communities, though, the linkage was direct, while in the case of the two North American communities it was indirect. In other words, in the United States and Canada it acted through, and together with, other variables that constitute one's social class. Despite widespread beliefs to the contrary, education per se is not associated with self-confidence in the United States or in Canada without the appropriate social class background and present class status as reflected in occupation and income.

It is interesting to examine another aspect here: the relation between self-confidence as an indicator of a sense of being adequate to participation in community social and civic affairs and actual participation itself. In the North American communities those who are high in this factor, that is, those with a strong feeling of adequacy or self-confidence and who have high socioeconomic positions are generally highly involved in community life. Another way of saying it is to recall that many of those who have high socioeconomic status are also high in self-confidence and these are the most highly involved people by far.

The importance of self-confidence for participation is underlined by another finding. Still speaking of the North American communities, those few who are high in self-confidence but are of lower socioeconomic status participate to a much greater extent than those of similarly low status who have less self-confidence. In the communities from the socialist countries the connection between one's presence in community events and this same self-confidence factor proved to be just as important as in the North American communities.

The situation of the two Italian communities turned out to be the most complex to analyze. On the one hand, the factor of self-confidence proved to be less closely connected to class variables and the variables of social background; on the other hand, the connection between self-confidence and participation was strong. What was different was the character of the latter relation. It was nearly nonexistent for the lowest social classes. They were self-confident relative to other classes there (or relative to people in other country samples), but they were at the same time very low in participation.

Put in another way, the two Italian communities have a greater number of persons with a high degree of self-confidence compared with the other communities. Consequently, one would have expected the highest number of participants there as well, given the participation-self-confidence relationship in every community. That, however, is not the case. The masses of lower-class Italians, even the highly self-confident, do not participate in civic affairs and even the highly educated self-confident Italians do not participate as much as might have been expected.

This Italian difference underlines the fact that to be successful in beginning innovation processes, it is necessary to take somewhat different actions and approaches not only within various political systems but also within individual countries.

The factor just examined, self-confidence, is not a result of biological characteristics. This emerges from what we have seen: that is, it's strongly connected in all of the communities to level of education. And it is also connected to sex.

At first sight a profound difference between the two sexes seems to exist. Women seem to have a very low degree of self-confidence. Further analysis, however, shows that the greatest differentiation in each community is not due to a person's sex but to socioeconomic role. In other words, the notable difference between men and women becomes modest when one compares men and women who are both working.

As for the connection between self-confidence and age, it is evident in all of the communities that self-confidence decreases or in any event is lower with age. And we think it is a matter of a personal loss of self-confidence with age rather than an historical difference of the last generation having had less self-confidence than the present one.

The interconnections among age, self-confidence and political involvement, are, however, not the same in the various communities. In the middle-aged group (from thirty-one to sixty) when the number of those who are self-confident increases, so does the proportion of those who are involved—in all communities. This is not true everywhere with regard to the youth. And it is true for the elderly in only two communities: in the North American communities where the elderly are also the least marginalized in comparison to their counterparts in the other communities.

Another dimension of personal perspectives was found, also by means of factor analysis, that we termed a feeling of citizen decisional competence. In this instance it was a matter of trying to find out if those interviewed were oriented toward opening the decision-making process to all of the people or if they believed that decisions should be reserved for specialists. Although this could be only a very crude indicator of a person's sense of his own competence, we could, because relatively few people in our samples were specialists, get an approximate idea by assuming the responses were projections in that regard. Did people believe that in order to be one of those who make decisions in the community, one had to have special preparation and competence which only an elite can have or that such decisions can be made by everyone, by the masses?

It is interesting to note that in the North American communities the fact of one's being inclined to opt for elite competence restricted to a few is not closely correlated to factors of class. It is true that as class position becomes higher, the number of elitists increases, but there are also large numbers at the upper levels who are oriented toward increasing the number participating in the decision-making process. People tended to be one or the other, elite- or mass-oriented there.

In contrast, in the Yugoslav communities (especially in the Bosnian community of Konjic) both attitudes were held simultaneously. Apparently the great majority intensely appreciate the requirements of specialization without thereby negating the competence of everyone in making decisions. In the two Italian communities the attitude structure turned out to be similar to that of the North American communities. In the Polish and Czechoslovak communities those interviewed were decisively oriented toward the possibility of everyone's participating in making decisions. The traditional socialist communities were more desirous of mass participation than were the capitalist communities.

It is, nevertheless, important to underline the point that, in general, appreciation for specialization is widespread in the communities of both systems. And it is a perspective evident at all levels of the class structure. It is not even closely connected to the level of education or to the other socioeconomic variables that have been connected to the other factors so far considered. This attests to the fact that the belief in the necessity of specialization as a prerequisite for competent decisional participation is by now universal in the modern world, including these socialist communities with the interesting, partial exception we noted in Yugoslavia. Such an attitude must be noted and understood as an obstacle to participation, especially if it is not at least accompanied by the dialectical feeling that everyone should participate, as is the case with the Yugoslav communities.

The Structure of Nonparticipation. To conclude this section on nonparticipation, I would like to attempt a synthesis of what has been described so far, a synthesis that will be continued in the next, concluding section of this chapter. We can say that we have outlined four dimensions of what, for the sake of clarity, we will call the model of nonparticipation. Following the order of my exposition and phrased in negative terms, they are: cynicism toward or pessimism about the representatives of local government; a feeling of impotence in influencing decisions concerning community problems; a feeling of inadequacy or lack of self-confidence; a feeling of participatory incompetence.

These factors are present in all of the communities, although in various forms and combinations. It is precisely this diversity from one community to another that provides the key for understanding more deeply present situations in any country. Let us review how these factors vary in the eight communities grouped according to the four models of participation outlined earlier.

Cynicism or pessimism is low in the Yugoslav socialist communities and almost as low in the North American capitalist ones. It is high in the communities of the traditional socialist model and even higher in the two Italian capitalist communities. In no community does such pessimism correlate with class variables, except in the United States community where the relation exists, although not strongly. There, cynicism or pessimism tends to vary with the indicators of social class along a curve that reaches its lowest point, that is, the highest degree of confidence in local representatives, in upper-class men and its highest point or lowest degree of confidence in local representatives in women with a low level of education.

The course of the feeling of being able to influence community decisions does not take a very different track. In the Yugoslav communities there is the highest degree of feeling able to influence decisions. And when this varies, as we have seen it do with regard to sex and occupation, in every such category the Yugoslavs in both communities feel more potent than their counterparts in all other communities. The North American communities follow, and not by much, but the difference by class is very great. The lowest-class people there, together with the lower-class counterparts in the two Italian communities, are the people with the most profound feeling of powerlessness.

One of the two Italian communities resembles in this regard the traditional socialist model with a very widespread feeling of powerlessness. The other is closer to the North American type. But since the feeling of being able to have such influence is not as high in the Italian communities, the variation by class turns out to be less significant in both Italian communities than in the North American ones.

Self-confidence or the feeling of personal adequacy, as we have called this factor, as it related to local political participation in the international study, proved to be of great importance for our understanding and development of future strategy. In both North American communities, this feeling of self-confidence was high, but only for the middle and upper classes. Moreover, an analysis (multiple regression, termed path analysis) revealed that this factor by itself affected the feeling of being able to influence community decisions. The importance of social class for the participatory situation in the North American communities was thereby strengthened since self-confidence was also a class phenomenon there.

The four socialist communities did not vary very much among themselves or in comparison with the other communities on their levels of self-confidence. In comparison with all the other communities they generally manifested low self-confidence, which was present, with few exceptions, in all social strata. The Czechoslovak and the Bosnian communities had a slightly higher degree of self-confidence in their citizenries than did the Polish and Slovenian communities but, we repeat, all were comparatively low.

In all communities and countries except for Poland there were strong and significant relationships discovered between self-confidence and local governmental/political participation. As an initial statement, then, we may say that the lack of self-confidence, a sense of personal wariness or caution, is everywhere an obstacle to participation.

But we can say more. In the two Yugoslav communities there were comparatively high levels of citizen participation although relatively low levels of feelings of personal adequacy. A closer look reveals that there was a tendency for the Yugoslav citizens of high self-confidence to participate more than equally self-confident citizens in the United States or Canada. And there was even a slight tendency in both Yugoslav communities for the citizens of the lowest levels of self-confidence to participate slightly more than equally cautious Americans and Canadians. (We remind the reader that we are using identical measures of self-confidence and that the Yugoslav citizenries are far less highly educated in terms of years of schooling, and much lower than the North Americans on the other social-strata indicators.)

Several major inferences can be drawn from these Yugoslav-North American findings. One is that while low personal self-confidence is indeed an obstacle to citizen participation, it does not have to be such a heavy obstacle as it is in the American or Canadian systems. Another is that self-confidence does vary by sociopolitical system and is not merely a product of natural, innate capabilities and experiences associated therewith in regard to success or failure in school. Still another is that, given what we know of prewar Yugoslavia, it appears possible to dramatically reshape in a relatively short time span the pattern of citizen participation and its dependence on a sense of personal adequacy. It is a safe historical reconstruction to imagine that few highly self-confident Yugoslavs of low levels of education or of manual occupations would have been participating much in their local community politics, except perhaps as revolutionaries, in the earlier periods.

The Italian findings, in fact, underline the point that the problem is not merely one of helping people gain or regain a sense of their own capabilities. In both Italian communities, a large number of persons evidence a high degree of self-confidence. These comparatively poor and poorly educated Italians in fact proved to be as self-confident as their much more affluent and educated North American middle- and upper-middle class fellows. And in both Italian communities there was also a connection, but very weak, between self-confidence and the feeling of being able to have a say in deciding how to solve community problems. There was also a much weaker association between this sense of self-confidence and socioeconomic class position in the Italian than in the North American communities.

The Italian lower classes especially, despite their subordination and often harsh exploitation, have maintained a sense of personal self-confidence that is astounding. But whether of lower-, medium- or upper-class position, Italians of high self-confidence in both communities participate much less in their local politics than do either the highly self-confident upper-middle class North Americans or the equally self-confident, significantly more working-class Yugoslavs. Despite the high degree of personal self-confidence, then, the Italians do not manifest their sense of adequacy by participating in a domain about which they are so cynical: local government and politics. It is apparently manifested instead in rich family, kinship and informal social relations. The major obstacle to citizen participation in local government/political affairs, then, is the sense of cynicism and of civic decisional impotency that breeds the Italian style of political alienation.

Indicators of Potential Participation

The preceding section was dedicated to an understanding of the empirical structure of, and the factors impeding, participation. Some elements were evident that on balance operate in one community or type of community negatively but constitute comparatively positive aspects in others: for example, trust in the representatives of local government. With respect especially to countries where citizen participation is now very slight, some of the questions to which we referred at the beginning will be reexamined.

Having seen the extent to which apathy and absence from civic affairs are widespread conditions of modern society in all of these systems except for the Yugoslav, can it make any sense to hypothesize that where we now find very little participation there are people who are generally interested in social problems (apart from those that touch them personally in a relatively dramatic way)? To respond to such a question we need to examine findings that appear to us as positive indicators of potential participation.

The Presence and Characterization of Interest in Social Problems. Data were also gathered on people's interest in having more information concerning the principal areas of civic life. This desire for more information was considered a significant sign, a demonstration of nonapathy, of the opposite of disinterestedness in community events, although not quite an indicator of readiness to participate.

The findings did not vary by variations in political system or groups of communities. Of the two participatory Yugoslav communities, only Konjic demonstrated notable interest in more information. The other community, Trzic, did not. The two North American communities also show different forms of behavior, although less contrasting than their Yugoslav counterparts. In this case it was the United States community that revealed more interest than the Canadian one. The Italian communities, instead, both manifested a high level of interest. The other two socialist communities, the Polish and Czechoslovak ones, were somewhere in the middle in this regard.

An examination of these data according to the variables used as indicators of class shows that the level of education is usually connected with the fact of one's being interested in having more information, but not in the two North American and the Polish communities. The lack of such a connection in the two North American communities needs to be kept in mind. It means, in fact, that even here where apathy seems to be widespread, lack of interest in civic affairs is not a generalized attitude nor is it pervasive in the lowest strata of the population, as might have been expected.

In a further examination based on demographic variables, it emerges that women did not evidence much less interest than men, except in the single case of the Czechoslovak community. In fact, in the Italian community of Scorzè the women turned out to be more interested in having greater information than were the men.

We recall that women, at least in our samples, generally have lower levels of education than men. Thus the connection between education and the desire for information is also less for women than for men. It appears, then, that the situation of the female populations with the one Czechoslovak exception is also more positive than is the masculine situation in regard to this indicator of potential participation; this is in dramatic contrast to the far worse participation situation of women.

In all of the communities the elderly showed the least interest in having more information. But it must be noted that their *relatively* negative situation in this regard in comparison with younger adults does not mean an *absolutely* negative situation. In fact almost everywhere more than half of the elderly responded positively; they did show interests in having more information about some aspect(s) of civic affairs.

As a further sign of interest in community life, people in all eight communities were asked about the changes that they held to be necessary for improving community life. Questions followed concerning their readiness to do something about such changes. Separate questions were posed about their readiness to give money, spend time or offer moral support, speak with their friends or try to influence others more actively in order to obtain the necessary changes.

Those in the two Italian communities were proportionately most change-oriented. They were followed by those from the Yugoslav community of Konjic and those from the Polish community. The people interviewed in the two North American communities gave fewer indications of changes considered necessary. Those from the Czechoslovak and Slovenian communities seemed the least dissatisfied.

It was somewhat surprising when confronted with this picture to learn that the two Italian samplings showed the least inclination to do anything in order to obtain the desired changes. But that was not at variance either with the slight participation or with the great pessimism or cynicism about the local officials mentioned earlier. A similar lack of readiness to act emerged from the Czechoslovak sample but with a different starting point, that is, from an expression of far fewer changes desired. A relatively high degree of readiness to do something is present in the North American and Polish communities.

The most interesting finding for various reasons is that of the Yugoslavian communities. Konjic evidenced the highest indication of readiness to act in order to obtain the changes considered necessary, which are many as we have noted. Trzic is not at an equal level but still manifests a generally positive readiness-to-act orientation toward the community, especially when one considers that there the changes felt to be necessary were far fewer than in the other communities.

For the Yugoslavs, then, the findings on readiness to participate are consistent with their participatory and other positive perspectives regarding their actually quite different communities with one, as we noted, being a developing industrial community and the other a more advanced one.

Precisely because of the meaning that these readiness-to-act attitudes have, it is worthwhile underlining the fact that the data on the two North American communities, when disaggregated according to social strata, reveal a characteristic that we consider positive. In the lower social strata, a substantial number of persons show a readiness to do something to bring about desired changes: thus, we find less pessimism or cynicism or even resignation than the paucity of actual participation, especially in the lower portions of the social structure, had suggested.

Some Hypotheses for Characterizing Initiatives for Participation. By putting together all of the elements that have emerged, we can outline some conclusions, which, at the same time, constitute starting points for developing a participation strategy.

With respect to participation four broad strata or types of people or perspectives can be identified:

1. those who are outside the happenings of community social life and apparently are not even interested in what happens there;
2. those who are outside such happenings but are interested in them, even if this interest seems to take the form more of wanting to be informed than of being ready to take part in community actions;
3. those who are outside but are interested and are also ready to participate in community events;
4. those who already participate, at least some of whom are also ready to become involved in a more major way.

In the light of such considerations, perhaps someone may suggest that the most opportune way to proceed is to give people more information and more education. It might be argued that this should constitute the ideal way to bring people step by step up the multilevel structures of nonparticipation, of potential participation and of participation found with different shapes everywhere.

The characteristics of persons now in these strata and the connections between and among the factors that inhibit participation and the socioeconomic and demographic variables suggest, however, another direction. That alternative is to concentrate directly and much more immediately on the goal to increase active participation.[16] Let us focus our attention on capitalist countries since it is there that we find ourselves committed to act. When we do that, we realize that there are several elements that are positive for the more direct path, granting the differences between the situation of the Italian and the North American communities.

In the Italian situation we have found an important element favorable to the possibility of directly opening up decision processes: the presence of classes and strata previously outside of all the important decisional positions in the key sectors of the community. They could constitute a natural political support for participatory initiatives.

An element that we can consider as at least an equivalent positive element in both North American communities is a more widespread optimism in the chances of bringing about change. As we have seen, people here were also more generally ready to participate in order to obtain the desired changes and that readiness was not restricted to the middle and upper classes.

In both the North American and Italian situations the examination of those who are ready to participate although not yet present in community activities suggests that a substantial increase in civic participation is possible with respect to women. To some extent, more in North America than Italy, this is also true with respect to the elderly.

This direct focus on initiatives of active participation involves consequences for the kinds of participatory initiatives that must be designed and carried out. We do not mean that better educational conditions and the greater availability of information cannot have positive effects on people. That is especially so if the content of both education and information is modified in order to conform to the objective of having people less disposed to passive consumerism and conformity.

The nodes to attack, however, if we wish to have more people present in decision-making processes are the following feelings: of inadequacy and of incompetence to take part in decisional processes; of cynicism or pessimism; of powerlessness in front of the power structure; of a lack of self-confidence. These hallmarks of modern society can be addressed directly and not only indirectly through an educational route. And our findings demonstrate that contrary to the popular image that increasing education is the easiest way to affect such changes, it is the most difficult. It is the most difficult because in the United States as in Canada educational experiences are heavily shaped and affected by social background, thereby creating great resistance to change.

Thus the initiatives to be taken must fundamentally respect two requirements. First they must open the decisional processes in such a way that those who participate in them can realize immediately how to act, how to be influential, and, second, they must be moments of egalitarian action that utilize the characteristics, experiences and knowledge that everyone has or has had rather than specialized knowledge or experience, which blocks right from the start people who try to participate and are thereby made to feel inadequate and incompetent.

Actually the foregoing analysis was based on a summary, composite kind of index that approximated the extent to which people declared they had been participating in three community domains: in local politics and government; in the local political economy; and in urban affairs. If we adjust the microscope so that we focus on each domain separately, the broader picture does not change in substance but clarifying details and differences do emerge.

In these three overlapping domains, the one in which citizens participate least is that of urban affairs. Even the major matters of employment and of attracting new industry nearly everywhere provide for opportunities for or the actuality of a bit more participation than do such urban affairs policy areas as public housing and urban planning and zoning. One might think that these more particularistic matters of urban affairs would have involved more citizens in communities in at least some countries, but in none is that the case.

When we look only at the most manifestly local governmental-political participation (in reference to matters of public finance and electoral-party-organizational representation of citizen interests), we find that the Italian aversion to politics and the Czechoslovak apathy constitute clearly distinctive patterns. The two Yugoslav communities are in first position, with the Poles not very different from the North Americans in intermediate participation positions.

In urban affairs, however, the situation is quite different. Here the Italian-American distinctions in citizen participation levels disappear. They are similar to each other and all are comparatively low. Only one-tenth of these four citizenries participate even once a month in such matters as urban planning, zoning or public housing. (In contrast, from more than one-fifth to nearly two-fifths participated to that degree or more frequently in local political-governmental affairs.) The differences between the two kinds of socialist situations are also substantially reduced in regard to citizen participation in urban affairs. The Yugoslavs are a bit in the lead but the differences between them and the more traditional Eastern European socialist communities are considerably reduced. In fact, the differences between citizens of the capitalist and socialist communities in regard to participation in local affairs is smallest in the domain of urban affairs and, we repeat, this is essentially due to the low levels of citizen participation everywhere.

It is in the matter of urban affairs, of urban policies and programs, that we undertake our detailed examination of the possibilities of opening a hitherto relatively closed institutional domain. More than local politics and government, more than the local economy or the local schools, urban affairs may take its place alongside health and medicine as domains controlled by exclusive, esoteric professionals. If, however, there is substantial hope of constructing programs and preparing initiatives for citizens actually participating effectively and en masse in urban affairs, as we hope to demonstrate, then there is hope that citizens may enter even more quickly and effectively into domains that will be less controlled by professionals.

We are engaged in the rest of the book in asking and trying to answer the question whether conditions exist now for the transformation of urban affairs from a closed, elite domain controlled by the few in capitalist and socialist countries today, even in the innovative self-managing Yugoslav system. In Chapter Five I will examine one initiative for such a transformation, the experience of preparing an urban plan to revitalize the Italian community of Faenza. Through an examination of what was done that was unconventional as well as through an examination of subsequent results, reactions and further potential for development, the reader may come to feel, as the writer does, that the answer to the question is cautiously positive.

Although we will deal with an experience that took place in an Italian community, an experience that followed and took advantage of the findings of the international study cited above, some of the dynamics seem easily transferable to the American situation. Such an experience, *mutatis mutandis*, can provide useful knowledge for those who, rather than functioning in the Italian sociopolitical context, move within the North American situation. First, however, it is necessary to take a long detour to consider the nature of the city and of urban policy-making, especially urban planning, from various points of view as well as from our own. Then we shall be in position to argue for beginning an opening in which citizens can participate in such a manner that it will not be distorting to speak of urban self-management.

2

THE CITY IN CONTEMPORARY
CAPITALIST SOCIETY

The Nature of Urban Problems

Due to the process of industrialization and the interest of power groups connected to industrial development, our societies seem to have become an "urban society." I am speaking now of the United States and also of the less developed urban society of Italy. This transformation goes beyond the merely demographic and physical sense of the definition, that is, beyond the increased numbers in "urban" areas and the extension of "urbanized space." Even the major social problems seem to have become "urban" problems. They appear to be the result of the kind of urban spatial aggregation characteristic of modern society.

Environmental pollution, the transportation crisis and the housing shortage appear to be urban problems as, in an even more dramatic form, do those of social fragmentation, anonymity, mental illness, neurosis, delinquency and violence. Given this situation, and despite the numerous analyses of the "crises" of the city and of society, the relationship between the two is insufficiently clear. From this lack of clarity, then, both the usual interpretations of the nature of urban problems and their solutions prove to be ambiguous, or actually wrong.

We begin here an examination of urban planning, the specialist's approach to the city, which has taken on a great importance in attempts to solve or reduce or even avoid urban problems. We will construct in a necessarily schematic way a frame of reference in which the first step will be to look into the nature of urban problems. We will define a model of the city-society relationship within the temporal space of the industrial epoch.

For the moment let us call the city a sociophysical system, that is, a human system that is actualized in space. The components of this system are persons who, rather than being taken as individuals, are considered in their interpersonal relations. The urban physical space is part of their lives and their relations in ways not yet sufficiently explored. As a result, we do not even possess the concepts and vocabulary to speak adequately of them.[1]

This is because the "fragmented and chopped analytical thought"[2] of the specialized disciplines of the modern epoch has taught us to see this urban space as a sum of various dimensions: the social (which includes the economic, cultural, etc.) and the physical (which is determined by the characteristics of natural and constructed space).

There really is, however, a unitary urban space which, for lack of a better word, we call sociophysical. This allows us to conceptualize more appropriately the city and society. The city is not merely the place or ensemble of places where society is located but the environment in which man realizes himself as a sociophysical being. The city is where most people pass the greater part of their lives and where their daily experiences take form.

Let us for the moment define the society as a system, as a hierarchical organization, with the reservation that this very basic assumption will later be modified and clarified. As such, it is a system whose elements are political, economic, social and cultural institutions that are organized hierarchically. The functioning of society depends upon their actions and their interactions.

Since society is realized in space, the actions of these institutions require physical localization. Thus, the institutions are present in various urban environments according to two differentiating coordinates: one determined by the nature of the institution itself, the other by the portion of the hierarchical structure present. In other words, perhaps in one city the top decision-making level of certain economic institutions is present while in others it is absent.

The relationship between the city and society is determined by the action of institutions. We can identify the city as the sociophysical ensemble of the human beings present in a given space, while institutions constitute the forms that condition people and differentiate life within that space.

The reality of classes in the modern city is expressed in the facts of institutional existence, especially the facts of hierarchy and exclusivity. Class membership is a result of, and at the same time determines, the possibilities of access to important institutions and to positions in their hierarchies. The dominant class is the one which controls the institutions by deciding to what extent they condition the ways of life of people belonging to other classes.

With these basic concepts in hand, we can now try to characterize the contemporary situation. We will do this by reviewing two dynamics from the start of the industrial epoch: that of institutions and that of urban, sociophysical reality springing from institutional interaction. From the viewpoint we are developing, two factors characterize the relatively few institutions of the beginning of the industrial epoch. One is the presence of a marked class homogeneity at the level of institutional decision-making, a homogeneity produced by the new urban bourgeoisie that held the industrial economic power. This class was, in fact, either directly or indirectly in control of the political, cultural and social institutions at the highest hierarchical levels both in the urban context and in society taken as a whole.

The second factor is directly derived from the connection of the institutions with the city as physical location, since in the first epoch of industrialism the city represented not only the place of production but also the center of other important interactions with other institutions. It also stood for the settling and locating place of that mass of citizens who made up the potential labor force and the market for the absorption of industrial products.

We can say that although in earlier historical time and space a complete coincidence between the city and society did not exist, there were many cities in which institutions responsible for the condition of the urban environment were situated. In other words, through time a significant amount of the power of society was located in specific urban environments, although in differentiated ways.[3]

The ongoing evolution of industrial society in the capitalistic environment is characterized by the forms that we can identify as belonging to a middle phase of development (after the first epoch of industrial capitalism) and to an advanced or so-called late industrial capitalism. These two phases succeeded each other in different ways in various nations, but within the scope of this synthesis that is not important.

The first of these two phases (that is, the middle phase of development) was characterized by the expansion of capitalist power groups in controlling production, finance and commerce in ever larger territorial environments. In the latter part of that phase and in the beginning of the second phase, that of advanced industrial capitalism, many such groups had an increasingly nationwide orientation.

Now that second, continuing phase is characterized by the international or multinational dimension of such power groups.[4] From our point of view, the important fact is that the middle phase accompanied and required the growth of particular cities in the Western and capitalist world, while in the present it no longer seems possible to single out from given urban or metropolitan realities a dependent relationship of the most powerful groups on the destiny of particular urban areas.[5]

This enlarging of the territorial range of institutional action, primarily but not only at the economic level, is substantiated by—and is being integrated with—other important processes.

First of all, class conflicts, which have characterized the development of society from the beginning of the industrial age, have resulted in the presence of representatives of traditionally dominated classes in the economic, political and other summits of society. This representation, however, has tended to be either token representation, which is *ipso facto* weak, or inauthentic representation in which the dominated class is represented by a person not a member of that class. Yet even this ineffective representation has not been operative in the significant domains of urban decision-making.

Although the dominant classes were being freed from the necessity of remaining in precise urban locations, this did not mean that they were no longer interested in the urban environment. On the contrary, it became ever more important as the place of the production and of the consumption of goods and services. The need to maintain, and even to enlarge, this condition led to the creation of ever more specialized institutions and new forms of conditioning through which extended forms of control over urban man's life occurred either openly or covertly.

The performance of the existing and the new institutions is characterized by highly developed and complex hierarchies. The conditioning forms given to the city, to all cities, come from organisms that are often foreign and removed from a given city or set of cities, even if still located in a fragmentary way in some of them. The pinnacles of these decisional hierarchies are, in other words, often elsewhere. If this holds true for economic institutions and private capital, it is not, indeed, less true for other institutional sectors and for public enterprises in capitalist and socialist countries alike.

The creation of public, as opposed to private, economic institutions in some nations has brought about the construction of forms of public control over economic development. Actually, these new nationwide institutions have become equally removed from local communities. They have tended to downgrade other institutions, especially those tied to precise territorial and urban environments, by helping to remove from these urban-linked institutions their decision-making capacity.[6]

Thus, the demographic and physical expansion of the city seems to have led, or is leading, to what has been called "urban society." This society seems to be represented only by cities and seems to have shaped itself into a generalized urbanized fabric homogeneously covering complete national and even supranational environments. At the same time, however, no improvement or reinforcement of the quality of the city is apparent. "Urban problems" have become its general condition.

Urban society, that is, represents in Henri Lefebvre's words, "an expansion of the city that destroys the city."[7] Better still, it gives witness to the destruction of the city, since the urbanized space of urban society becomes an anonymous, flattened, sociophysical space lacking nodes with the ability of self-direction and self-definition. This can be seen in the physical images of the multiplying suburbs or in the spread of megalopolis.

This city-society relation within "urban society" allows us to outline the nature and the source of urban problems, which actually do not derive from the city even if they are manifest there. Urban problems derive from the presence of specialized institutions within the largest dimensions of society. One can understand this better from the other perspective already mentioned in the evolution of the institutions-city relation: the perspective of the industrial city.

From its first appearance, industrial organization was characterized by a clearer separation than had occurred before in the life of an individual between the moment of work and the moment of private life. *Homo economicus* becomes increasingly separated from the human being as a whole. More and more precise, specialized and partial roles are his. The problem of his self-determination in the environment and the philosophical-ideological perspectives of the early industrial age became one of control over his job. Doing a job is his most important capacity and also absorbs much of the everyday and existential time of his life.

It is exactly to facilitate this control that the physical separation between one's place of work and one's living place becomes important. Studies on the initial years of the industrialization process, in fact, have hypothesized that it was not the technical requirements intrinsic to the production processes that concentrated the new industrial activities into specialized places, producing the so-called "factory system." It was instead the result of efforts to control human activity, the work rhythms of man-as-machine. The separation between the life of factory work, which is subject to the rules and conditioning of the "world of work," and the rest of everyday life was extended to other sectors of economic activity. All the other types of work involved in mass production, including office work, became sectoralized and separated from the other domains of life. In fact, the same process led to the development of the pervasive modern tertiary economic organization. "Workers living by the pen or typewriter" were organized into and controlled within the typical factories of the economic or governmental bureaucracy: the office buildings.[8]

With regard to urban space, the crucial point in the logic of industrial development occurred when land passed from being considered as use value to its present exchange values. Urban property that is controlled by economic institutions is used according to the amount of profit that can be gotten from it.[9] This consideration of land as exchange value, which characterizes the early years of the industrial age, marks the moment in which physical space began to be treated apart from its social connotations. From then on, the composite sociophysical connotations of urban space, which nevertheless continue to be its essence, are no longer treated as such in the specialized disciplines and professions. In other words, although its holistic essence remains, urban space is considered analytically in pieces and as pieces.

The development of such disciplines as urban planning was—and still is—determined by the definitions of the problems and the terms with which the problems were expressed during the evolution of industrial society. These problems range from the location of industries within particular urban areas to those concerning the living conditions of the urban masses. The description of urban planning as the "discipline that confronts urban problems" is obviously the ideological rendering of the dominant class, and it becomes the instrument for addressing the needs and objectives of this same class.

Later on, in the advanced industrial epoch, when urban planning reached its more definite disciplinary formation, this became clear from its very pronouncements. In fact, its objectives were sharply synthesized in Corbusier's famous and profoundly influential *Athens Charter*, which marked a fundamental step in the evolution of this discipline some forty years ago. They were:

· to assure mankind of sound and healthy lodging;
· to organize places of work [so that] work will once more regain its character as a natural human activity;
· to set up the facilities necessary to the sound use of free time;
· to establish links between these different organizations by means of a traffic network that provides the necessary connections while respecting the prerogatives of each element.

The charter continues, "These four functions. . .are the four keys to urbanism. . . . [10]

For a deeper analysis of the operations of this organizational form of the city of specialized parts, the mapping involved being technically called "zoning," I refer the reader to other writings.[11] I will return to this later on because it is a characteristic still present today in urban planning operations and consecrated by legislation in many countries. Moreover, functionally segregated cities and zoning seem to be the substance of proposals that are still considered to be avant-garde as in the case of organizational criteria of superurban areas.[12]

The organization of the productive sector based on the division of labor has, according to Karl Marx's analysis, led man to those conditions of alienation that marked the industrial age in its capitalistic forms. Such alienation was present when Marx made his analysis, but the important fact here is that the same partializing, segregating approach to urban organization, and not merely in physical zoning terms, testifies to a new type of alienation, or better put, to an increase of alienation, due this time to further divisions in the heart of urban sociophysical life.

These divisions are no longer rooted only in the divisions of labor but also in the multiform divisions that life has undergone. Urban man is fragmented and atomized not only in his work roles but also in the roles he assumes in his so-called social life. He is alienated not only from his economic products but also from his cultural creations and social relations. Admitting for the moment the utility of making such distinctions, nonetheless insofar as they are distinguished by separations and boundaries, they are artificial and artificially created.

Indeed, as I have said earlier, the dominant class had sought the separation, and the separate organization, of "work" from "life," because it was particularly interested in controlling the productive side of urban man. Consequently, the development of industrial society into its modern, consumer-society forms focused attention on this consumer aspect of man's life, until then considered residual and only important with respect to the good performance of man-as-worker. The change occurred for at least two reasons. An interest in controlling man as consumer developed, since in his role as consumer he is potentially capable of affecting or even controlling production. Furthermore, the development of class conflicts made it important to avoid in the developing urban concentrations the kinds of uprisings that had occurred in the factories with trade union formation. It was even possible to recover through urban consumption (and production) the power that had been lost in the factories to the newly organized.[13]

It is in this way, then, that the above urban model, with an unorganized citizenry ever fragmented further by zoning, is valuable. It is grounded on, and legitimizes, a model of urban man as atomized. This urban man, inherent in the urban model that generated him, is controlled in all of his parts, functions and roles by specialized and partialized institutions. This is valuable precisely to those who control those institutions.

In fact, it is due to exactly this initial dissociation between public life (the life of work) and private life (home life, free time, and so on), imposed through the economically powerful institutions of the early industrial age, that the proliferating mechanisms of manipulation, exploitation and class control have developed as they have. These mechanisms, operating in a consumer society, are based on such divisions of life.[14]

To the person who must accept the hierarchically ordered life of work and respect its roles and rhythms is offered the "freedom" of self-expression in private life as well as a measure of consumer goods and services. This freedom, however, tends to become increasingly fictitious and trivial since it is increasingly shaped by, programmed for and made dependent on decisions coming "from outside" of the individual's life and of his circle of intimate and close relationships.

Man undergoes this conditioning in a largely passive way since he no longer faces it as a whole being, as a social being in the totality of his relationships. He progressively becomes more atomized and more isolated. The result of this conditioning and the forms of life that it imposes is the progressive disintegration of man's social relations, an increasing disinterest in life, a sense of unreality and apathy toward the happenings of social life. In this light and in such an urban society, man feels more and more impotent in exerting any kind of influence.

Thus, the urban environment, in which the division of labor and the specialization of functions and roles are maximized, as is the division of life in all its daily expressions, becomes the place where man as social being is most alienated. This alienation can be described as double-barreled. Since the person *qua* person is no longer recognized, there is no longer any space for him to express himself in a total way. He can only define himself in places or in decisional processes if he is a member of relevant institutions. And then he can only express partial and specialized opinions that will carry institutional weight not because they derive from a human being but because they derive from his role and his status within the institution to which he belongs.

Thus man as a social human being enters into a process or a condition in which he feels himself to be less than a total person. He is barely able, and over a period of time becomes less able, to control or to determine the expressions of his life. He is an object made up of parts that function together and which require servicing in order for them to perform. The search for the appropriate services is carried out not in the name of man's totality but in the name of his memberships and his various roles in institutions which afford him the right to such services.

Thus, urban problems, ranging from those dealing with physical space (problems that become manifest in the use of this space) to so-called social problems, derive from forms and conditions in which the recognition of man's humanity in everyday urban life is denied. In this urbanized space man finds himself increasingly alienated and estranged from other human beings, even to the point of being alienated from himself.

The historic city itself disappeared exactly when it seemed most exalted, in the form of "urban society." Sociophysical urban space has always been the place and ensemble of places in which the population in every historical period (but for different reasons and with different forms) has concentrated and lived out its social relations, its civic life, as a collective reality in order to address and resolve its collective problems. The disappearance of this city, then, can be explained by the fact that its sociophysical space has become, or tends to become, the encounter points for apathetic and atomized people who have suffered or have accepted the condition of not being able to administer their own affairs nor address their common civic problems. Those problems originate from, and are controlled by, forms of power that exist generally at a supracity level and have a different logic from that of preurban society.

Urban institutions in and for their development demanded an ever greater centralization of human beings. They did not, however, need the traditional sociophysical dimensions of the city in its active and creative forms and therefore in its autonomous and self-determining forms. These modern institutions required agglomerations that were simply sets or sums of individuals. The suburbs, the surrounding towns of large cities, the anonymous and socially flattened but sprawling megalopolis, are forms that correspond to this requirement. It is in this direction that urban development has been pushed. Thus outlined, the nature of urban problems can be traced to a model made up of two relations, which, in turn, are interrelated.

1. By strengthening the institutional fabric, the power structure has created a condition in which the members of the subordinate classes are subject to increasing fragmentation not only within the productive system but also within their urban social life and within their very human essence. (The institutional-development logic is external to the urban environment itself to which the power structure is increasingly less bound.)
2. The multiformed and progressive alienation of people, within the context of class stratification and institutional differentiation, underlies and shapes the problems "of the city" and the impoverishment of its sociophysical space. (Lefebvre again: "Urban alienation becomes the container, the substance and the perpetuation of all the other forms of alienation.")[15]

Given this characterization of the nature of urban problems, their resolution can only be effected by a fundamental transformation of modern society. Such an option must involve and comprehend man's de-alienation as an urban social being. This process in turn can only proceed according to two interrelated principles:

1. Man must recapture his totality as a social human being by

overcoming the fragmentation and partialization to which he is subjected.

2. Man must gain decision-making power within urban social life, demonstrating thereby that he partakes, in an active and critical way, in the processes of society.

This hypothesis, which shapes the following pages of the book, calls for further specifications. There is a major Marxist formulation that reaches similar conclusions but takes a different route to do so. In fact, the concept of the "city-factory," the organization of territory integral to the logic and the needs of the dominant class, is similar to the hypothesis I have formulated but it seems to me that there are two substantial differences. That Marxist analysis treats the concept of economic production as distinctive and presumably primary. In our view such a conception, though widespread and not limited to Marxists, is very wrong. That analysis also leads to a strategy that declares its protagonists to be suprahuman organizations rather than human beings, organizations that remain suprahuman even if they are constituted by social classes, and even if lower-class individuals legitimize them so that they might actually be representative.

Moreover, that analysis ignores the problem of the fragmented state of those who belong to the subordinate classes, their apathetic state resulting from their institutional partialization and, often, marginalization. It is precisely those members of the subordinate classes who are most subject to marginalization in its most widespread, if not its deepest, forms.[16]

A rather different Marxist hypothesis is based on the objective of reconquering the human being's "right to the city," meaning the reconquering of the various nonalienating human relationships within the forms of man's everyday life. But what forces and tactics can be proposed to bring about this de-alienation? How can this process be started?

These certainly are the questions which we must address. One can begin by examining the current activity and development of society, which can provide either partial or total answers. Furthermore, one can begin by seeing and understanding those elements of contemporary society that might constitute its fulcrum points, and those principal elements of contradiction that might be used to start these processes of change. But we know that these processes must aim at producing an environment for urban life in which nonalienated human beings become the subjects of experiences. In other words, they must be able to participate as total persons in making decisions that concern them as total persons, rather than as people whose particular specialized aspects are affected by specialized decisions.

There are many proposals and initiatives circulating currently that seem to be, or succeed in being, considered definitive solutions to urban problems (or at least to some of them). Most of the proposals are efforts to rationalize present urban society by modifying the most unbalanced features of the urban environment. Such approaches historically have tended to subdivide urban problems and to separate them by singling out specialized solutions to them. In very recent formulations the accent has been put on a "systems" vision of the problems in view of their interrelations in the urban environment and in society. Even in this more complex vision, the attempt to eliminate urban problems is connected to the possible perfecting of cognitive tools used within specialized disciplines or by the "systems scientists" themselves.

Another attitude proposes instead that urban problems can be solved only if power relationships *within* economic and political institutions of society are modified. Or, it may be asserted, they can be solved if the forces of the traditionally subordinate classes take control of the economic and political institutions of society. Such attitudes all presuppose the maintenance of such institutions in their fundamental character.

In such frames a particular role is entrusted to economic planning and, in the urban environment, to urban planning as instruments by which institutions can put an end to injustices. They can put an end to a certain type of laissez faire that has characterized the management of bourgeois society, which, presumably, in its lack of concern has been the basis of both society's and the city's social, economic and other problems.

The solution to urban problems and the transformation of society, however, cannot result from a different operation of presently existing institutions. This is made clear by planning operations, including plans that are inspired precisely by the forces of the left in such countries as Italy, where the left now controls the bulk of local government and urban planning. It is in the light of the above erroneous considerations and the unhappy consequences that might result for the marginal and dominated groups represented by parties of the left that I will examine in the following chapter the nature of an urban plan, its functioning and results in the situation of today.

Such a task has an ulterior motive: that of constructing an example to show how a specialized operation functions. Such an example, with small variations, can serve to illustrate not only urban planning but also the other forms of planning presently used by institutions in the exercise of their power, with due regard for their particular organizational forms.

The fourth chapter of this book will seek to specify some concepts that can be used to define the objectives of the transformation of society. Such a task is given meaning by elements deriving from reality that validate the possibility of a plan of action that starts from the urban environment but is also based on the active presence of people. Another basic hypothesis here is that the process of de-alienation cannot be carried out through representative delegations or the political institutions of various classes; it must be accomplished by people themselves acting in opened and therefore transformed institutions.

The Technocratic Approach to Urban Problems and the Concepts of the Left

The technocratic approach to urban problems arose with the development of industrial society in which the city was divided into special areas and functional sectors. This model led to a perspective according to which urban problems were seen and understood as problems of a specific area and/or a specific sector. The disciplines supposed to solve them have become progressively more specialized and specific. Since the problems of the various sectors were and are in turn subdivided and studied separately, the development of deep but narrow disciplines was natural. This approach has marked (and still does) the development of capitalistic society and can, above all, be characterized by its contribution to the compartmentalization and sectoralization of urban life. In other words, problems are confronted separately and the "solutions" proposed tend to produce situations of ever greater divisions.

What is proposed and carried out in one sector is done so according to the logic of that sector and usually is not related to what is occurring in other sectors or areas. Furthermore, the very procedure of specialized disciplines directs the search for ways to solve problems toward the technocratic forms produced by experts. The general attitude toward the problems of the city is that which grew up along with the development of "problem-solving" methodologies:

every urban problem can be faced as the search for the best technical solution.

The distinctive character that this approach evidences is the lack of will or capacity to face urban problems as they really are: as problems with much wider boundaries and of whole persons. By being considered separately, urban problems are viewed above all, or only, according to their superficial manifestations and not according to their real causes.

This is justified by the "philosophy" of the specialist called in to contribute to the solving of a specific problem that is formulated and proposed in a certain way. Therefore, he considers that as the context requiring his attention, the context in which he must demonstrate his expertise. Corresponding to the specialization is the fact that those who hold the power and are aware of these problems in their wholeness usually have no intention of facing them in their essence. This would mean to unhinge or to put an end to the mechanisms of exploitation and advantage that, in many cases, operate precisely in the situations that are rooted in these problems.

Both old and recent analyses have verified this fact with respect to the so-called "housing problem." The scarcity of adequate accommodations is not a matter of technical solutions but one of group interests.[17] The urban transportation problem provides an even more obvious example. This sector has attracted attention for some time now: from the dramatic problems of urban traffic congestion and chaos to the difficulties that commuters experience in going back and forth between their homes and their jobs. Recently these problems have become more widespread, but even decades ago they were acutely present in various large cities of the West.

There were two basic sensible attitudes toward such problems. One was to organize territorial development appropriately. This meant the development and location of productive activities (not only industrial) in such a way as not to provoke explosive expansion of certain urban areas while emptying others; not to expand some productive sectors and destroy others equally good or better—or at least deserving of survival. The other intelligent attitude was to try to reduce the thrust toward a near monopoly of private transportation and to seek solutions to the needs of territorial mobility not in an individualistic and narrow but in a responsible public and broad way.

There were of course some who warned against the move to the automobile, to private ownership of them, to "auto-mobility," seeing in these machines something fatal for the quality of urban social life, for the economy and efficiency of urban transportation.[18] But this was not the attitude of most.

One of the first results of the traffic problem was to produce "traffic specialists." It is the task of these technicians, by means of specific instruments and methodologies, to "provide solutions" to these problems, conceived as essentially autonomous traffic problems even if their autonomy is only apparent and fictitious. What resulted was the production of extensive studies and the development of ever more advanced and refined technical "solutions," a treating of symptoms and not of causes.

As the traffic problem illustrates, one of the characteristics that generally emerges from the technical-specialist approach to urban problems is that it provides a kind of first-aid solution to external aspects of problems. These so-called solutions in turn generate larger problems and the need for even more expertise. They require still more ability, knowledge and the use of "special" instruments.

But let's return to the traffic problem and its real causes. They could and can only be rationalized and receive superficial treatment as long as one remains within a system that tends continuously to expand suburbia and to extend urbanized space into continuous megalopolis. Modern man is "spontaneously" led to choose private means of transportation just as in another sector he is led to choose his small family house in an ever more distant and dull suburb.

These so-called choices are made on the basis of social organization that alienates just because a person is not given an opportunity to choose on any but the predetermined individualized basis the mode of meeting a problem. He can only satisfy, in a private way, those needs to which he has been educated to read as "his." The traffic specialist only reinforces an unduly narrow and distorted definition of such needs.

A vicious circle is thus created. By accepting these conditions, the alienated individual acts on his environment in such a way as to provoke further problems that only increase his fragmentation, privatization, isolation and, thus, his alienation. In fact, his lack of choices in the realm of public goods and services is often an epiphenomenon and always consistent with his lack of choices in other, nongovernmental domains marked by institutional hierarchies and skewed distributions of power for which nonfree markets do not compensate.

The inadequacy of the technical-specialist approach to urban problems is related to the ideological-philosophical perspective of early industrial society. In the more advanced industrial society of today an attempt has been made to adapt the approach to more modern understandings. I am referring here to the systems approach to urban problems. This approach does not deal with single problems but with interrelated groups. It views the city as a system with the properties of a system. Furthermore, it is viewed as a system in which sequences of events are determined by the laws of probability.[19]

This type of approach is apparently sophisticated and general. By interpreting the city as a whole system, it seems to overcome the barriers that separate the various disciplines. Its real characteristic, its true nature, however, takes the form of a superspecialization, a sort of second-degree specialization. It has already led to the creation of new roles: systems specialists, specialists in the construction of models, software and hardware designers, programmers. These new specialized forms, in fact, respond to, attempt to respond to, the requirements for more widespread and in-depth control over urban and human life in contemporary advanced society.

This can be seen by examining the most significant result brought about by the urban systems approach: that of constructing "models" of the city. Such models are built by a relatively extended series of mathematical relations that in the most advanced models are made up of linear, nonlinear and differential equations. They allow for the representation of complex situations characterized by the temporal variation of selected conditions.

The construction of models only functional through the use of an electronic computer is intended to place those who make decisions for the city in a situation in which they can base their initiatives not only on the intuitively formulated proposals of traditional experts, of technicians and/or administrators, but also on plans (software computer programs) evaluated by means of computer simulation processes. From the reactions produced in the model, one can then judge the positive or negative qualities of such proposals in the context of the city as a general system.

Certain aspects of this approach help us to understand how it has a logic of an ever more advanced technocracy. First of all, the expert or group of experts, in constructing their model of the city, construct relations that must be expressed quantitatively in order for them to be mathematically functional. Given the many qualitative aspects of urban reality, therefore, they choose those that will be the "surrogates" of the qualitative aspects inasmuch as they can be expressed in quantifiable coefficients. The experts themselves decide which ones they will include and which they will ignore.

These operations are made up of subjective evaluations, which, however, are described in terms of objective operations and/or neutral scientific analysis. Such purported value-free operations are supposed to be exempt from the conditioning that derives from personal and sociopolitical value systems.[20]

The presumed scientific nature of the model makes the expert alone qualified to bring about modifications of it and to decide which are the most opportune, that is, which do not destroy its essential traits. There are those in the ranks of scientists and technicians as well as some politicians who believe that the best hope for common citizens is in the further development and use of such advanced scientific tools.[21] But the actual situation is very different.

The new technicians, the superexperts, elaborate their models on the basis of society as it is presently or as it is understood by them, these superspecialized persons, as well as by the persons who may have commissioned the model. They see the city and society as always more complex systems, the operating needs of which the superexperts derive from their vision of complex fragmented beings, not from men. The complexity that they manipulate is addressed to the search for coexistence, for a nonexcessive, manageable or tolerable degree of conflict among the institutional or subinstitutional parts. This has constituted a subtle, often unconscious reinforcement of the conception of these commonly accepted parts as existing, as real, when they are merely fictitious abstractions.

The methodologies and the proposed application of systems theory to urban planning seem to remove citizens from the condition of having to submit to the irrational and intuitive choices of past urban-planning specialists and to open up larger possibilities of citizen control and citizen intervention. In effect, however, this occurs only subsequently to enclose citizens within the walls built on those new attributes of rationality and, above all, of objectivity and scienticity.[22]

Indeed, the still "intuitive" and "irrational" choices of the urban planner, of the urban specialist, at times can be understood and seen as inherent in his work although the rational-scientific mantle of the new technocratic superspecializations tends to camouflage such realities.

If for no other reasons than these increasing aspects of specialization and the increasing distance of persons who live and interact in a given urban milieu from the decisional processes relevant to it, we can recognize that these new approaches, theories and instruments are not such as to contribute effectively to the solution of urban problems. But our concern with them here has an additional basis. They constitute a tendency in modern society that has just begun: the attempt to absorb the city into a sort of ecumenical vision. In other terms, there are efforts made to have the city disappear for good by dissolving it into a sort of uniform mush of urbanized space more or less extended over the entire milieu of the national and/or multi-national system's control.

Apart from the more successful science-fiction images by which this ecumenical vision can be defined, as in the works of Buckminster Fuller or Doxiadis, this is not a way of strengthening the social qualities of the city.[23] Nor is it a diffusion of the positive sociophysical aspects of the acclaimed "city effect," at least in the forms of contemporary urbanized society.

As various writers on so-called "post-industrial" society suggest, however, such homogenizing happenings are likely, however unfortunately, to characterize and also condition the future development of many modern urban societies.[24]

The superspecialized disciplines and the experts, in constructing and operating increasingly complex models seemingly under more effective control, become a necessity of the system. They become a necessity precisely to the extent that in the dominant classes these same experts become an ever more important system-guidance or conflict-management component, and, for that very reason, ever more influential.[25] In fact, a broader control of the physical extension of cities is needed, but one that is profoundly different in character from this technocratic approach.

Returning to the city and to urban problems, we can draw the following conclusion. Neither of the two aforementioned approaches proves adequate; that is, neither the technical-specialist approach, the one dealing with partial and sectoral problems, which continues to be the more widespread form of action, nor the systems approach that tends to connect these partial matters by putting them into a programmatic bundle, often with the backing and sanction of internationally powerful groups.[26]

We may call all those planning perspectives that are premised on the assumption of maintaining the foundations of the present capitalist system the "capitalist planning approach." As we have suggested, we find there a tendency toward an increase in urban problems continuing until urban problems apparently disappear because the city itself will have disappeared. The city is increasingly dissolving into an anonymous, uniformly urbanized space where most men and women live in an ever more controlled and alienated state.

Webber's "urban non-place realm"[27] is destined for the subordinate classes made up of the traditional working class and of those classes belonging to the middle and small bourgeoisie of the past phases of industrial society. They are all impoverished and proletarianized not in terms of their material well-being but in the apparent disappearance of their participatory desires and the actual impossibility of their active presence in the decision-making system. Power-holding groups because they are increasingly served by technologically more advanced means of communication and transportation, tend to build even more isolated and physically less concentrated and visible "citadels" for residential and economic-political purposes.[28]

In short, we can understand that the lines of urban action deriving from those who hold top power in the capitalist world are oriented toward the maintenance of such power by means of an increasingly centralized and technocratic management of urban processes and problems whether directed to growth or stagnation. Other ideologies and forces are, however, present on the world's urban social scene: do alternative perspectives regarding the problems and development of the city come from them?

By reviewing the lines of urban development in industrial and capitalistic society, in fact, ideological elements emerge that were born from a socialist perspective of the urban situation and from a primary concern with the powerless rather than with the powerful. In general terms, it is perhaps not wrong to say that interest in the city has constituted until recently an enduring but auxiliary concern of Marxist thought and writings. The economic system has been given the dominant strategic and theoretical role, especially the system of industrial production within which the dominant class perpetrated the most evident exploitation of the working class. But now where socialist regimes have come to power, urban development and control become matters of much more priority although the city-economic nexus is still as powerful as ever.

The interest in urban life of Marx, Engels and subsequent Marxists resulted from the often inhuman conditions in which the industrial society's urban masses lived. In these conditions a further form of the exploitation of the working class was recognized. The division between the economic system and the social system, deriving from the organizational and ideological-philosophic forms of industrial society, led the working class itself and its representatives to differentiate its working world and urban life demands. The struggle was directed against the same antagonists because, in the famous phrase, government was merely the executive committee of the bourgeoisie. On the one hand, though, it became an effort to win a different relationship of power within the place of work, principally the factory. On the other hand, the struggle took the form of requests for services within the residentially separated places of social life.

It is precisely the concept of the "right" to have such facilities, termed services, which has been characteristic of the entire history of urban planning and is one of the hallmarks of present planning, that constitutes a hollow victory for the poor and the less well-to-do in the urban milieu. The right to services is essentially a socialist vision even though it now is widespread.

We must clarify this critique of services. The acceptance of the division between the (productive) economic system and the (nonproductive) social system had a double-pronged consequence. First the expropriation of other forms of surplus value in the domain of social life was not recognized as such. Then man's needs began to be put only in the form of requests to obtain "services," which services were often primarily visible and evaluated in terms of economic costs deriving from their use. Benefits, especially noneconomic benefits, could not be calculated and often seemed to cease to exist.

Services became significant for the satisfaction of diverse forms of partialized needs, both those which the development of industrial society made evident (for example, a minimum level of formal education for everyone) and those which the new society had exacerbated (for example, the requirements of mobility in going back and forth between one's residence and one's place of work).

Nearly all of these various types of needs were first responded to privately insofar as they constituted forms of profit for those in the position to sell such services and privileges for those who could buy them. Only gradually did they become public responsibilities. Having lost their profitability on a mass supply basis,[29] many of these services were not attractive to private businessmen and became the business of public government.

In view of the progressive conquest of ever new services, one might wonder whether a dialectical dynamic of system-maintenance and of system-conservation was present. On the one hand, the working class won victories and, on the other hand, it was in the interest of the dominant class to have these services absorbed by the public domain. For the dominant class this meant constructing a net of specialized institutional structures with which to satisfy the needs of the subordinate classes and, thus, to control them by providing sectoralized and fragmented forms of these needs and need-satisfying services in hierarchically organized modes.

Such partialization is in addition to the partializing effects of the division of labor itself, a fundamental ingredient in the development of modern industrial capitalist and now also of socialist society.

It is this very aspect which is the most important: the inherent meaning and role of services, even when they become public services, in the life of urban man whether in capitalist or socialist societies. The mechanism for the creation of services has functioned by following the logic of segmented man that the ideology of bourgeois society had originally imposed but which socialists also accepted—at least until government withers away and utopia is reached.

That logic held that given a human need requiring servicing, a specialized, separate institution was charged, or if necessary created, to take into account that aspect of man. In this way an operation was carried out that certainly seemed most functional and efficient. This strategy, however, has very negative results, at least as it has developed historically. This is the case, above all, for those of the subordinate classes who are, in fact, obliged to submit to the public service structures. In the process, and perhaps worst of all, the human being's need for a vital human experience, a need integral to his entire person, was depersonalized and dehumanized and treated by a segmented and segmenting bureaucratic institution.

Keeping in mind the fact that the expression of a human need as an experiential action is also an act of production, this need and the related action come within the ambit of a corresponding bureaucratic organization that monopolizes the power in such production-and-exchange relations. In fact, those who alienate are also among the alienated. Why? Because those who provide or deliver services ignore, however unwittingly, what distinguishes human needs from those that are, in general, animal needs; in no case and under no conditions are *human* needs constituted only of partial and material elements as occurs in the matrix of servicing of human needs.

In other words, a process has been put into motion. The human being has been taught to identify, subdivide and categorize his needs according to the particular services organized and furnished by the same society. The service itself is a moment of passive acceptance of the satisfying of a partialized need. That in turn becomes a relevant factor in the further fragmentation and partialization of needs.

All the so-called services are evident examples of this state of affairs. Much public transportation is obviously meant for the subordinate classes. The objective, however, never was that of creating transportation conditions for the human being so that he might in some way find in traveling an occasion to meet and exchange pleasantly if not creatively with other human beings, to have total human experiences. Instead, the objective has always been organizing an expedient operation of temporary storage of people on anything that moves in order to transport them from point A to point B. This is so despite that fact that some people succeed in making such services more social, more human, and even if first class services are potentially more pleasant although potentially more hierarchical in their human relations than second or third.

It is easy to believe that if this same need, rather than being managed by a specialized (even when public) institution, had remained within the competence of those persons needing to travel about, they might have found more adequate forms. Their response might have taken into account an image of themselves as total persons rather than as quasi-inanimate objects to be moved or transported. At least it might have if the entire specialization scenario had not developed as it has.

Nor does the functioning of the "service" spare the worker who furnishes it within the frame of the specialized institution. He works, as we remarked, in the same conditions of alienation as the recipient of the service. He cannot determine the forms in which he provides it, nor can he control the effects of his labor. He cannot participate as a total person but can only exercise that side of himself that is specific and restricted to the role he fills in the institution. We say "cannot" knowing full well that, fortunately, a few people manage to maintain their humanity and their humane relations even while performing servicing operations to or on people regarded as somehow deficient.

In the domain of urban problems, then, and especially with this concept of public services, the left generally, in Europe as in North America, appears to sustain the division of labor and of life that for Marx was one of the basic aspects of a capitalist conception of society and from which he indicated the crucial need to be free before passing to a communist society.[30]

There are now various contributions to the criticism of specialized institutions in the administering of public services.[31] The most interesting proposals come from precisely those who suggest the opening up of institutions in their three key features: that of their functional specialization; that of their hierarchical form of management and that of their exclusivity, that is, the requirements of special and particular prerequisites in order for one to be an institutional insider.

Despite the increase of criticism, however, the promise of adequate services remains one of the pervasive principles of the other basic concept that has been diffused throughout society: that of the necessity of public planning. The latter also has come from the left but is found throughout the political center by now.

In this regard the two apparently different patterns of European countries and North America, especially the United States, are perhaps interesting. In European capitalist countries a method of direct management by the government was demanded by those who considered such public action as an alternative to the philosophy of free enterprise. Gradually, as the position of the left became influential within national and local political structures, public planning grew ever stronger as it assumed the character of a method for transforming society.

The "publicly planned" development of the city was in European countries the goal for those who were outside the power structure. They gradually succeeded in making the dominant groups accept such a direction or, more recently, they themselves acted at the municipal level as they became a majority force in controlling these local structures of public power in such countries as Italy. The concepts, models and instruments for such planning, however, were forged within the capitalist situation in total consistency with the ideology that shaped bourgeois society.

I will return to this aspect later when I examine in depth how, from the principles of bourgeois society, the urban plan became primarily addressed to physical characteristics, and especially to economic implications and consequences of land uses. Its specialized function is assigned to the jurisdiction of the local government under the control of political institutions operating at other levels. As a specialized operation it is based on specialized operators, urban planners, who are entrusted with the plan's preparation.

Clearly, in no European country were urban plans per se seen as *the* instrument for transforming society. This is also true today. But the emphasis placed on them by forces on the left, especially after the war, was based on two assumptions: first, that urban plans were effective forms in helping to transform society. By taking care of urban and territorial problems, political consensus could be built by and around the forces on the left. Through planning specific control over capitalist exploitation that focused on speculative income deriving from use of the land could be instituted. The other assumption was that urban plans provided instruments that would permit the new socialist order to be transferred to the forms of urban life. This was to occur once power was won within the general political system.

The American planning process was and is different in various aspects. In fact, despite a much more general consensus opposed to planning, a current in favor of public planning was finally generated. This occurred within that ideology that might be termed liberal, the connotation being privately oriented rather than publicly oriented or socialist. Naturally, seen from this perspective, public planning has had to guarantee that it will attain bourgeois ideals otherwise not attainable, a not so easy task.

As Ernest Erber writes in the introduction to *Urban Planning in Transition*:

> City planners occupied themselves with design for improved civic appearance, with traffic movement, parks and boulevards, subdivision of land and zoning, location of schools and town halls. To the extent that their efforts collided with political machines that made development subject to political needs rather than to a master plan, planning became identified with municipal reform, usually a business- and taxpayer-based movement.
>
> Within the pluralist spectrum, therefore, planners became aligned with middle-class, business, home-owning, largely suburban forces, because this was the only segment of American society that had need for planning, if only in a severely restricted role.[32]

Suburbia and small towns around the centers of large cities, outer cities, became the proplanning protagonists. For the advocates of urban planning for big cities it was a question of bringing order to increasingly more chaotic and more decayed inner cities. This meant the physical renovation and social "cleaning up" of slum areas, the arrangement of modern traffic networks, adequate parking space, the building of parks, churches and other facilities useful in giving the renewed city an ordered and "respectable" appearance.[33] Both business and financial groups plus upper-middle-class people saw urban planning as a vital and important tool in such urban redevelopment activities.

From an even more altruistic and optimistic liberal viewpoint, planning was conceived as an ensemble of more substantial actions meant for the improvement of urban life. In this milieu the emphasis was placed on a kind of planning leading and related to huge direct public as well as massive private expenditures in the area of housing and services.[34] The emphasis of all those American planners, representatives of these needs for transformation, was increasingly placed on plans as the instrument for obtaining greater social justice without losing sight of, and often through, stimulation of the local economy.

Given the failures of the New Frontier and the Great Society, as well as that of the subsequent "benign neglect," and with urban problems and crises mounting, the view developed that the urban planner, through his plans, should take the part of the poor, the disinherited and the powerless. In such perspectives the planner is to work to bring about conditions for the simultaneous improvement of housing, transportation and other services in an urban situation especially for the disadvantaged, concentrated most often in the cores of the inner cities or in backwaters of the industrial society such as the mountains and hill country of Appalachia.

Conditions in the largest American cities have worsened considerably in recent years as their socioeconomic fabric decays. The bare survival of many smaller cities has further increased the number and concern of those who support the need for more public urban planning, despite the seeming increase in, and vocal opposition by, opponents.

Although the socioeconomic situation of North America is different from that of the various European nations, with further differences between Canada and the United States, planning and the plan represent here as well as there a method and an instrument for attacking urban ills by improving, and also to a certain extent by modifying, the socioeconomic-political situation.

The question that must then be raised is whether or not urban planning and its instruments truly constitute nodal points for the transformation of society, for progressively confronting and alleviating urban ills.

The reader will have noticed that while coming very close to a definition of the city I did not quite give one. Although I shall not engage in an extended consideration, the subject must receive a bit more attention here. This is so not only because our society seems to be increasingly an urban society rather than a society of cities. It is also because something so seemingly self-evident as the existence of cities may, with a little reflection, prove to be not evident at all. In fact, cities, if they ever existed in North America, with a few early exceptions as in the colonial period in the United States or as Quebec City in French Canada, may not have existed for a very long time. They became cases of arrested development or of atrophy even before the development of the modern urban and suburban population agglomerations. Rather than being a continent or nation of sick cities, as the conventional widom suggests, North America (particularly in the United States with its melting pot a single vessel of uninational character compared with the more binational and multicultural ethnic mosaic of Canada) is a continent of countries with, at most, very underdeveloped cities.

What is meant by referring to the United States as an underdeveloped country in respect to its cities when everyone knows America led the world in urbanization? Let us see what we have said about the meaning of a city so far. We said that the city was a sociophysical ensemble of the human beings present in a given space, in which space is also a set of institutions such that, through time, particular and variable amounts of power of society are localized. Institutions are themselves people, subsets of people organized in usually exclusive and hierarchical ways to carry forward projects of concern. The more a place is a city, the more concentrated or dense is the power, including the power, the potency, of many if not most citizens in at least some major domains of so-called everyday life.

A missing ingredient in this conception or definition of the city is this: cities were places wherein relatively advanced divisions of labor and of life (relative to more rural countrysides or agricultural areas) fit together in such a manner that people within the city boundary as well as some immediately outside found and felt their life together, their life in *this* bounded space, was meaningful and distinctive. Their particular residential location was meaningful to them, however vaguely conceived were the distinctions they made of their citizenship compared with citizenship in other cities or outside of any city. The mix of institutions had a crosscutting dimension such that members of the city community, residents and/or citizens, participated in various life-fulfilling or need-satisfying activities *as* members *of the city*, as total people in a rich everyday life.

The pattern of fitting together did not necessarily mean, indeed, rarely meant, a harmonious conjunction of parts or a pleasing form. The fitting together might have been and often was due to the pressing domination of one group, category, class or subset of people. Or it might even have been in the hands of a conqueror. At times a city might have been a place of comparative peace and quiet and at other times conflict, rapid change, even internal bloody strife. Participation in the life of the city was merely participation with people but with people who had a sense of their rootedness relative to past reality or present desires in the *particular* city whether or not that sentiment extended to a sense of being a simple representative or mighty ruler of that place and its people.[35]

This is not the time or the place to say in detail why, but it appears to me that the most accurate reading of American history is really not one of cities in financial crisis today but of the earlier failure of towns and urban places to have become cities. Formally they are cities, of course. They have governments, of a sort, of traditionally very limited power as creatures of the larger state governments which, in their turn, compared to private institutional pools of power, were relatively restricted institutions. The American historical experience is precisely one of limited government compared to European situations. And it is one of incredible geographical mobility pursuing economic opportunities always less restricted to particular cities or spaces. Sam Bass Warner, Jr., describes as a fundamental aspect of Philadelphia's development the perspective and practice of privatism.[36] Combined with a profound, deep privatism is a process of suburbanization and, despite growth, a constant removal from cities that started in colonial times and that is not yet finished. The conclusion we reach is that the city, as such, is and has nearly always been a fragment of what it is supposed or thought to be in the United States of America.[37]

There is an equally strong and equally faulty opposite image of Italy as an ancient nation of cities; of a highly cultured civilization resting on the roots of the city of Rome. It is all that except it is not an ancient nation. Italy is a newer nation than is the United States. It was not a nation until the time of the American civil war. It was a complex of small city-states, states often with a dominant central city but usually with several distinct and distinctive cities. Whether the dominant civilization was Greek or Roman, French or Spanish, or a mixture of several including Arabian and even earlier the Etruscans, cities had profound and distinctive meanings for their citizens. Traces of such meaning remain even today in the specific personal character and personality not only in regions but of particular cities. Even after the walls that protected and help to define the lines between city citizens and "strangers" were overpassed or became merely internal center boundaries, Italian cities like many European cities still had a meaning, a social-psychological significance for people that was manifested in a plethora of small and larger ways.

Paradoxically, however, with national unification came a movement away from local government and power. Instead of replacing locally dominant noble families or the Church with locally based and usually locally visioned bourgeoisie, the unitary government centered in Rome made the city a creature of central government ministries and functionaries. Urban planning became a matter of centralized state direction and permissions, until rather recently when control began to be returned, mostly theoretically so far, toward the lower levels of regions, individual cities and sets of associated small communities. In other words, while Italian cities in particular, but also those of some other European countries, had a much more variegated regionalized and localized economy compared to the more unitary national economy in the United States, the former had less of an opportunity to strengthen its localistic city fabric because of the nature of the unitary state. The relatively recent post-World War II mobility and construction of a seminational economy in the north of Italy as well as the introduction of national television and the like has resulted in a movement of Italian cities toward what we may term the American pattern of cities or noncities. And of course with the invention of the multinational corporation as well as of the Common Market the historically more particularized and localized European economies are moving toward the transcity American pattern.

Conversely, in the United States as well as Canada there was more decentralized state/provincial and municipal rule. Despite the onset of some degrees of city home rule, however, the so-called city has remained nearly impotent as a civic-citizen reality. True, the industry of local government and politics exists, but even that is an industry of national scale. City managers and city planners move from place to place and coast to coast and professional politicians often move as well but up toward state, regional and national centers of power and away from cities. In the twentieth century as the urban places called cities in the United States acquired tools such as urban planning and zoning, beginning pretty much in the twenties, those places were to witness great increases in the waves of suburbanization and more recently of small-town and city growth outside of metropolitan areas, and a consolidation of the American national state that kept cities retarded in their growth and development—as *cities*. Contrary to the title of an award-winning book by American political scientists, *The Civic Culture*, the United States is not a country of a civic culture.[38] In fact, the relatively low participation level there, compared with other countries, despite the comparatively high levels of material goods and services, of various services and even of education, is not unrelated to the near absence of cities and of truly civic cultures.

We have said all of this about American cities because it is important to keep in mind in the following discussion what urban planners and urban planning are really concerned with currently. Whether city planners or urban planners working for other units of government, the American planner is not working with quite the same phenomenon that his Italian counterpart is working with. Despite the differences in the reality, in the existential reality, of city life in America and Europe, however, the planner and the planning process are more similar than different. Perhaps this is because we are reaching a point in history where a more universal homogeneity is the fate of mankind whether we like it or not. Our entire argument, though, rests on the assumption that such is neither desirable nor fated. It is neither if urban planners, with others, understand what it is they do rather than what it is they are said to do—as well as what it is they may begin to do. We may now start to answer the basic question raised earlier about the possibility of urban planning and its instruments constituting nodal points for social transformation by beginning to examine urban planning and its instruments.

3

URBAN PLANS IN CONTEMPORARY
CAPITALIST SOCIETY

The Nature of Urban Planning and Plans

Since the time when civilization first took on an urban character, urban planning existed. One can identify its early characteristics as the spontaneous or programmed form with which man had organized the transformation of natural into urban or urbanized space.

Whoever held the power controlled the urban space and used it according to his needs. These ranged from constructing military defenses to constructions that simultaneously served the power holder's comfort and prestige. The rest of the space was simply shaped by the everyday life of the common people who were, of course, controlled by those constituting the dominant group. Normally, though, this control left ample freedom, within contexts of historically forged customs and traditions, precisely in the expressions of daily life. It affected particularly the more strictly economic and military aspects that seemed fundamental for maintaining the conditions and relations of security and dominance.

When we speak of "modern urban planning," however, we know we are referring not to its earliest periods when it was integrated into community life. An Italian definition also good for North America is that urban planning is the "scientific discipline of the organization of urban space."[1] Since urban planning has become an institutionalized "discipline," its very nature has been modified. For this very reason, it is no longer integrated with the other forms of social life. It is separate and entrusted to particular persons and specific activities.

During the course of modern urban planning, major disputes have centered around its contents and role. An echo of the debates between those who wanted to concentrate on the problems dealing with the creation of "urban form," usually in architectural/physical design terms, and those who, instead, considered it fundamental that urban planning be concerned primarily with the economic aspects and functionality of the city still exists.

The former maintained, and still do maintain, that changes in urban society may follow changes in urban forms. Those who hold this position specify that "this process is not linear. . .the forms act on human behavior only through feed-back. . . ."[2] One can also find in this position those who propose to seek the redemption of the contemporary city's often miserable conditions in the "beauty" deriving from the designing of artistic urban-architectonic forms.[3]

There are others who instead see the organization of urban physical space in terms of land uses. They tend to believe that the balanced development of the territory or the recuperation of the present situation of the city must rest directly upon the economic health or recovery of the city. Planning should thus deal with regulating and locating functions and services and these also are to be controlled by the principle of economic maximization.

The debates over these "visions" of urban planning, which are periodically very intense, are in my view tangential. This is so whether the dispute concerns the autonomy of architecture from urban planning, or vice versa,[4] or whether it deals with the so-called interdisciplinary realm insofar as it becomes a battle for the right of economists or urban planners to perform the primary role in determining practical goals. In other words, these debates have not really contributed to an understanding of the nature of urban planning. By "understanding" I mean its reinterpretation, its critical evaluation, its eventual redefinition in the light of the problems and needs of today's society for the society of tomorrow.

Thus, in seeking to contribute a critical rereading of the nature of urban planning as it has operated in its traditional-discipline form, let us begin by considering it in terms that nearly everybody agrees with. Modern urban planning has become a discipline for controlling the use of physical space. In other words, urban planning affects or controls the ways in which human beings interact with physical space—or tries to do so—in theory and in practice.

Some may basically agree with this definition and yet find it too restrictive. In fact, different situations exist in different countries, and varying situations also exist within the same countries. With respect to some situations more than others, for example, one can speak about theoretical (or pure) and applied urban planning. But, in my opinion, in these countries, certainly including the United States, "practical" decisions about the use of physical space are kept far from urban planners. Thus, urban planning theory, to say it very brutally, is an expression of the impotence of "applied urban planning."

This is clearly the meaning, whether intended exactly in those terms or not, of Erber's passage following the one already cited:

> Despite the limitations placed on planning at the local level by too little authority and territorial limitations, and perhaps in part because of these constraints, planners applied themselves assiduously to devising a methodology that would make the practice of planning subject to the recognized principles, standards and techniques common to a profession. How else could they master the knowledge for planning in a basically anti-planning culture other than by selfdevelopment?[5]

Or as an unsparing critic of American urban planning, Frances Fox Piven, herself a planner, said of the professional planning beliefs of the fifties:

> City planners, we were taught, were the rational facilitators of urban development. It was our special role to assess the needs and goals of the city over time, to survey relevant action alternatives in the areas of land use and physical development, and to assess the future impact of these alternative development strategies on community goals.[6]

In regard to "the decisions that shaped the form of our cities" she went on to say, correctly, we believe:

> The key decisions. . . were never decisions embodied in any plans made by planners. Compared with the formative influence of capital investment decisions, planners and their plans were mere shadow play. At most, planners only struggled to service the cities built by private capital with the support of public capital.

Things are apparently, but not substantially, different in Canada. There, public planning seems to have had more weight in overall urban development. Actually, however, this is not so if the authors of *Subject to Approval*, a recent review of municipal planning in Ontario, are correct: "Few are the planners, at effective levels of power and authority, who have devoted more than a fraction of their time, energy and technical resources to anything beyond the regulation of private development and formulation of public service programs required to support and accommodate such development."[7]

Perhaps the difference is that in Canada, alongside this limited presence of public planning in urban development, planners (within the universities or various research institutes) did not even take refuge in the theoretical study of the discipline. Nor would this have represented a better state of affairs.

It is the widespread conception of the autonomy of theory and praxis that, in my judgment, needs to be discussed with respect to urban planning and not only with respect to it alone. In fact, it does not make sense to conceive of a "theoretical" approach, thought or model-building that is germane to or relevant for a city unless it also is a "theor-act," that is, being simultaneously an action addressed to so-called practical problems or functional praxis.

As a result, in urban planning the proposal to organize physical space by means of making plans has no sense only as a pragmatic action; it is meaningless or barren if it is not also a theoretical reflection. The latter becomes, in turn, a pure academic exercise if not understood and treated as action.[8]

If we begin from already formulated theoretical conceptions of the city and urbanized space, we see that they entail visions of organization and of practical actions to remedy organizational problems that may occur. As David Harvey notes in his *Social Justice and the City*, the city is seen by some as primarily an ensemble of humanly built things organized in space according to certain models.[9] This, then, provides for practical problem-solving addressed to patterns of a built environment. The preparation of a certain kind of planning vision of "emergent evolution" in the urban world, like that of Doxiadis, leads him to formulate plans of "spectacular mystic-design."

So far I have spoken about proposals dealing with the organization of physical space and about plans in a nearly equivalent way. In fact, plans are more complete and precise formulations to the extent that they are instruments of organizations that are authorized and guaranteed by local government or are under the aegis of the state itself.

Thus, in seeking to understand the nature of urban planning as a traditional profession or academic discipline, it is necessary to understand the plan. The plan is the most developed instrument by which urban planning achieves its major objective: the organization of space. The first important consideration emerges precisely from the nature of plans. Here one can see that many of the disciplinary debates of the past, or those now in course, dealt or deal with pseudo-problems. Generally, in fact, in the dissertations on the definition of urban planning, the aspect which has been emphasized and underlined the most, the one considered determining, is the "object" of the discipline, the noun "physical space." It seems to me that the real weight should be given to the verb, to the action of organizing physical space beyond which hovers the acting subject, that is, the man.

Modern urban planning was not shaped into a "discipline" by accident, given the occurrence of two simultaneous phenomena. With the advent of industrialization, the power structure gradually assumed a more articulate and complex form. On the other hand, the problems of social organization began to connote "urban problems." In *The City in History* Lewis Mumford underlines the transformation of the land from use value to economic and commercial value as the most important change for the city. As he emphasizes, this meant that the model of city development was no longer only related to human needs and activities and to the needs of power; it was also related to the needs of commercialization of land.[10] This modification also became the prelude to a dichotomy that was more fictitious than real: public versus private interests in the use of physical space. It was fictitious because whether planning was expressed through public authorities or privately, the plans invariably ended by representing the interests of the dominant class as well as this class's interpretation of the needs and interests of the other classes.

In any event, it was this apparent contrast between public and private interests especially that led to the institutionalization of the controlled use of physical space through urban planning.

The plan does not have its special significance in forecasting a specific use of physical space but in making the forecast a requirement. Its nature is that of a juridical instrument the norms and requirements of which avail themselves of graphic representations rather than written directions for the sake of convenience.[11]

The two points that emerge from what has just been said can be summarized as follows:

1. Man is the referent of urban planning, which is involved on his behalf in dealing with (regardless of how) forms of life in space. Space in itself is not the referent.

2. Modern urban planning's involvement has assumed the form of plans and instruments for proposing, directing and binding. These are always instruments of control.

If this is the nature of urban planning, it is relevant at this point to see precisely how urban planning acts on man and further, for what kind of man it acts. There are two extreme alternatives. One, the importance of urban planning in modern society and for today's individuals is very little or none. In other words, its nature and its instruments in effect deal with aspects of life that are secondary or trivial. Or two, urban planning is an important determinant in the present development of society and persons.

If our analysis led us to conclude that the first of these two alternatives is true, we would no longer overly concern ourselves with defining the nature of urban planning. We would stop asking why it is not effective in facing up to the problems of the city and we would probably decide to spend our energies in another way.

If, instead, we concluded that it is important and influential, we would have to become concerned with its methods and instruments. We would have to see seriously what is wrong and how it can be changed.

If we wish to observe the real state of affairs, we must, first of all, understand that in reality men, while theoretically "equal," are in practice profoundly different. Further, we are examining the reality of Western society. Therefore, we know that social classes exist and that there are class differences whether or not one prefers a word other than class. A question thus suggests itself. Is urban planning equal for all? We have seen before that it is an instrument and that for all its potentiality to unfold, it must be a juridical instrument, something like a law that is guaranteed by the power of the state.

As our first objective then, let us see how the plan functions in relation to the life of people. In other words, let us try to see if the mechanism of the plan itself can produce justice and what kind of justice this might be.[12]

An urban plan is a prevision of how the development of a city or territory should take shape in a given future. As such, and like all plans, not merely urban ones, the plan functions as a "reducer" of the variety of the possible uses of physical space.

In this regard Stafford Beer in *Designing Freedom* explains that institutions need a series of these variety reducers. All the major institutions that regulate society are high variety systems, that is, they have a large number of possible states. All of them, however, even if highly adaptable and flexible, have limited times within which they can reach a balance after a change from one state to another. But the continual disturbances to which the system is subjected tend, instead, to work against that. Such dynamics, if not dealt with, can lead to system rigidities or even collapse. Speaking of such institutions as systems, Beer comments: "How do they cope? There is only one way to cope, and all institutions use it—although they use it in many forms. They have to reduce the variety of the system."[13]

The plan, then, is one of these reducers within the urban social system.

We may now return to the preceding point. The social system is not a homogeneous whole. How does this variety reducer function with respect to different social classes? Three possible situations can be noted.

The first concerns those who have no control over the plan and its operations. They can only accept what the plan prescribes. Often they do not even know that the new conditioning factors of their lives come from and have been legalized by something called a "plan."

The second is the position of those who have power and influence at the local level. Thus, they can intervene in the very process of the plan or can seek to control that which they define as good for themselves and their conception of the entire community.

Finally, there are those who are above the plan for two kinds of reasons. Their power is of the highest level but, above all, their interests are far reaching and not affected very much by the forms of land use in a given area of physical space. Thus, they are indifferent to the plan's process. They assume that in case of need they have enough power, if necessary, to modify a law stemming from a plan or the plan itself. Or, at most, they can pay what is for them a relatively low price to avoid such an obstacle.

It is easy enough to identify three broad social strata to which these three situations pertain, but it is perhaps still easier to illustrate with everyday examples. The first example is that of a family which has for some time lived in a building in a given outlying area. The building is without any historical relevance, but also not modern although in more or less decent condition. The family pays a modest although actually excessive rent for the house, but it is not as much as rents for new houses. Let us suppose that the plan foresees for whatever reasons a high density construction for that area.

This family does not know that there even is a plan nor does it know what a master plan means. Yet, it finds itself dispossessed. The outcome may be relatively good (that is, with some notice or even a cash payment) or bad (that is, recourse to lawyers and expensive although futile lawsuits). This situation occurs because the owner naturally finds it convenient either directly to build a new building according to the size and the characteristics anticipated by the plan or to sell it to someone else who will do so.

It matters little that in that area this family has much more than a dwelling to inhabit; the family must give up friends and daily relationships, and life habits that have become significant social relations. As renters or tenants, this family can do little or nothing to forestall the event. This kind of situation is by now so frequently repeated, obviously in a totally "legal" way, that it has come to be found quite normal.[14]

The second example concerns a company of local and national importance with interests in a particular area (among other things, it has a large number of workers and employees). It also owns a rather extended area that is a central point relative to the various parts of the urban settlement. The area is currently not used, so the local government decides that it is the ideal moment to set up facilities there necessary for public transport services.

The plan is made or changed with this zonal use in mind. The company finds that this use is not profitable for its future, having programmed the development there of a much more remunerative residential subdivision. Thus, it begins to exert pressure to modify that part of the plan. The local government changes the plan for that area.[15]

The third example concerns a company of national scope that wishes to build an oil refinery. Within certain limits, it is not confined to a precise location or a specific municipality. It has some locational choice. It consults with various town and city and county governments.

It seems as if it is impossible to find a good solution to its needs. All the municipalities reject it because of such negative consequences as pollution, the anticipated great needs of water, and the heavy traffic that the installation of a refinery brings.

Obviously, the story finishes in the usual way. If it is not located in one place, it is located in another. Someone from such a company almost always succeeds in being persuasive enough, since he has the means to be so. The municipality that finally accepts the refinery already had a plan that obviously did not foresee the installment of a refinery. It was basing its previsions of a "balanced development" of the community perhaps on the improvement of agricultural resources and tourism, since it was part of an area especially qualified for both of these economic sectors.

This municipality has clearly set aside all of its forecasts, has modified the intended use of its areas, has inserted new facilities and effected all the necessary changes. The large industry, of crucial importance on the national level, was satisfied. No matter that this came or will come about in a way damaging for many. Aside from whether or not its locational choice was necessary or correct, the installation is located to the detriment of the appropriate use of that community's resources. As a whole, the community evidently is less powerful than the large public company of top-level national importance, or at least many citizens of such a community are.[16]

But there is another element in the nature of the plan that affects the social equity of it, independently of the contents. In its present institutionalization as a juridical tool and as a "reducer of variety" within the social system, the plan is by its nature different from the social system to which it is applied. Contrary to the social system, the plan is neither self-varying nor self-adaptive; it is quite static and rigid even though it may be changed or amended.

Consequently, a contradiction occurs. On the one hand, in its role as "reducer of variety" the plan must itself be rigid, with precise contours and contents. On the other hand, however, it must be flexible in order to meet the nature of a system that is constantly changing. Any unexpected and unforeseen change that the social system itself succeeds in experiencing or absorbing must threaten to invalidate immediately the plan in its rigidities.

As a result, in order to compensate for the different nature of the plan and of the social system, there must be a compromise. The plan must have some marginal flexibilities and ambiguities. These must serve to allow changes without arriving at frequent and bureaucratically difficult changes in the plan. This flexibility and vagueness, obviously, spell greater degrees of freedom. Thus, once again the presence of social classes with different power and different economic possibilities in society works so that this greater freedom benefits certain groups and classes much more than others.

Indeed, it is the well-to-do classes that have most of the relevant information and knowledge or can acquire it through consultants, professionals and experts. They can thus better know those degrees of freedom and at the same time have the economic means for using them, if and when they are interested in so doing.

Thus, a primary conclusion is that urban plans control and condition a person or a group in inverse proportion to the level of his or its status, wealth and power. This sheds further light on the various approaches I am examining.

In fact, urban plans, insofar as they are institutionalized and are part of a capitalistic social structure, are not neutral instruments dependent merely on the intentions of urban planners and administrators. Consequently, these plans are not likely to help in bringing about social justice even when they are used by urban planners on behalf of the poor. This leads us to another interesting observation.

Let us take a look at the situation of local administrations in most countries within the capitalistic system. They are roughly differentiated into three types. The first is the type that we can call conservative, in which the interests that express and control the local administrations are clearly identifiable with those of the traditional dominant classes. The second is the liberal democratic type, in which the local administrations do not intentionally take into account the class structure of society and do proclaim the theoretical ideals of egalitarianism. The third is the type that we may label progressive, oriented toward the "promotion"' of the working and subordinated categories and classes and toward the amelioration or even transformation of the socioeconomic structure.

Excluding some extreme examples of the first type, that is, characterized by unrestrained and brutal speculation, one cannot generally say that the use of planning instruments by various local administrations of these different types results in major differences, at least from the perspective of their physical or land-use development. We cannot say that even progressive administrations have created cities or parts of them that can be clearly distinguished from those of the others in terms of having a different quality. This is so even if for some time now the progressive administrations have had functioning plans and/or have been able to block the most brutal speculation and/or have provided various sport, health and educational services available to all.

This is not due so much to the situation of the capitalist system in which these local administrations function—we will see how the city in socialist countries helps to demonstrate this—but derives rather from two series of factors that reinforce each other. On the one side, the criteria and methods used in traditional urban planning do not result in urban quality. At the same time, they interfere only tangentially with those factors in industrial and capitalist society that degrade everyday life and, for that very reason, urban space—that is, urban sociophysical space.[17]

If this primary conclusion has answered the question concerning the influence that urban plans can have on society and social classes, other questions arise that can help to give an even more adequate answer. If the plan is not egalitarian in its mechanisms, what of its contents? To what point are these in turn meaningful, given the reality of class in capitalist society?

Once again, I believe that the best way to answer is to examine first what happens theoretically, that is, what plans are supposed to be when formed according to the accepted principles of the academic discipline of urban planning. Then we shall examine in reality who makes the plan, who operates the plan, who commissions it, what the plan's objectives and actual effects are in practice.

As a first step, let us see what methods of urban planning are used in preparing to make a plan, how real people are taken into account and then how they are influenced by the plan.

Models and Methods of Urban Planning

Let us examine step by step how the construction of an urban plan occurs, meaning the "plan" par excellence, the master plan. Since urban planning no longer produces manuals as it once did, the "lay person" would find it difficult to get clear explanations. Even if there is a paucity of official manuals, however, this does not mean that the traditional tools have been seriously called into question or that urban planning is going through a crisis. Perhaps this was and is the case for some. It is a fact, however, that in most universities where urban planning is taught, the way students are taught to produce a plan has not changed basically for many years. This is so in professional practice as well, at least in the great majority of cases.

We present a description of what this process is supposed to be. When commissioned to do a plan, the urban planner begins his work knowing that he must accomplish certain basic operations: the gathering of data on population and economic activities and the collection of data on the use of physical space (that is, on the location of public services, of the principal economic activities, and on the residential fabric). Other specific studies, such as a historical study of the urban settlement and growth patterns or particular traffic studies, may also be performed. For the moment, though, I will leave these special studies aside while speaking of the more central activity.

Once the urban planner has obtained the data and information, he evaluates them. More precisely, he carries out two operations: that of quantification and that of location. The first involves all those counting and arithmetic operations concerning such things as space for services, residential expansion, the extension of productive activities and of the use or reuse of the actual building stock. The second includes all those decisions concerning the location of public services but also of other functions, activities and facilities.

Thus, the urban planner superimposes the knowledge and theory which make up his professional baggage and experience on the often "raw" data and information he has gathered. He preoccupies himself with the "when," "where" and, finally, the "how" of locating activities and functions that characterize the desired urban human life in his community or jurisdiction.

He presupposes that this means one must organize or reorganize the city, conceived as the physical space already urbanized or the space which will have to be urbanized in the future in order to meet the needs of order and public well-being. Said in another way, the planning procedure should guarantee that the plan's prescriptions be a way of regulating the interaction between man and physical space so as to maximize two aspects: *the quality of human life in urban space and the equality of opportunity to use this quality of physical space.*

With the traditional way of drawing up a plan and its optimal theoretical objectives thus established, it is now necessary to see this process, and what in fact happens, from a more detailed perspective.

Data-gathering

Data usually are first gathered on the total population of the city at the time in which the study is made. To this is added data on the historical variations of the population over a relatively extended preceding period.

I am pausing over this first phase because it has a particular importance. Indeed, the clever urban-planner expert feels he can already use this data as an important source of information and evaluation. He is already able to place the city in one of his predictive formulae: the city of X to N-times-X number of inhabitants with an increase of population in the last Y years of a given average quantity Z per year will. . . ; or, the city of X to N-times-X number of inhabitants with a population that remains nearly constant in the last Y years will. . . and so on.

These categories do not only function analytically for him. On the basis of such classifications, a standard description is derived, a standard evaluation. Consequently, the urban planner already begins to predict probable solutions to the problems that he presumes or knows he will find.

This procedure, normal in the human mind, proves to be particularly dangerous in drawing up a plan. This is so because the situation of the cities is infinitely more complex than the few categories that the urban planner has preestablished on the basis of some numerical parameters. Thus, he risks (for the present I'll only say "risks") falling into the errors that one usually commits when proceeding to such simplistic categorizations. These errors involve thinking that one can associate qualitatively different behaviors through classifications derived from numerical or quantitative variables.

According to traditional methods that have become nearly intuitive parameters for us, we almost instinctively attribute to quantitative data indicating the population expansion of a city the index value of that city's prosperity. Actually, in most cases this growth instead is accompanied by a rise of problems in which it becomes difficult to judge whether or not that population increase is a positive or negative factor. It would also be equally wrong to consider every negative variation of the numerical value of the population to be a negative indication, or vice versa. In fact, population losses in London or in various American cities are often taken as a negative fact, as a warning of their decline and degradation.

It is evident, though, that the urban planner does not stop with such general quantitative data. These, however, I repeat, help him to come to very early conclusions that are successively verified or qualified by means of similar procedures, that is, the conversion of certain quantitative data into qualitative indices. The relation between the toal number of those who are employed and the number of those employed in industry, or an index deriving from the relation between the total population and the employed population are typical examples. They are considered important indicators of the "economic health" of an area (and not only for the purposes of planning). Such indices are among those considered fundamental in classifying the extent to which the area of a certain region or nation should be further developed.[18]

In such cases as well, the simple knowledge of these numerical relations is not enough to understand what are the real economic conditions of the population. Nor is it enough to help one to picture the problems the people may have in that city.[19]

In making a plan, the urban planner is also interested in such other numerical values as the distribution of the population by age and sex, its occupational conditions, its family dimensions, housing conditions and housing statuses. He often derives this information from the official census which may not be able adequately to represent rapidly changing situations.

This, however, does not seem to me to be a major problem. Indeed, it is partially or totally overcome when and if the local administration is able and wants to organize its own supplementary data-gathering and statistical services. In such cases the urban planner has a source of fresher and more precise data than from the census. The important point here is that by gathering such data, the planner engages in a grand, erroneous operation. It is erroneous because it is subdivided data (the data mentioned earlier or other additional data); it is conceived and collected partialized category by partialized category. In doing so, the planner loses in the process the very possibility of encountering the persons or groups of persons from whom the data come. At the same time, the planner believes and makes others believe that he is obtaining a profound understanding of the social fabric he is investigating.

In fact, once he has gathered the data on his list, he uses it in substantially limited and misleading ways. He works on each sector separately without even attempting to reconstruct the needs, the necessities and the desires of persons or social groups that cut across or are beneath such separated sectors. He treats single aspects separately and constructs on them the inputs for services and functions, rather than dealing with holistic, experiential needs of whole people.

Let us take an example. If and when the urban planner is engaged in determining educational-facility needs, he works with such data as the age of the population. From that he estimates how many future pupils can be expected, usually without considering if they will be from small urban or big rural families or from poor, middle-class or rich upper-class families. Even if all this would be important information in improving the "educational services," the planner does not operate in such a "sociological" or "political" manner.

Another limited and misleading way the planner uses such data concerns the fact that he must make decisions of another kind. Those decisions require the putting together of the data of more than one of the preceding categories. How can he do this with his categories of census data? He cannot really say who among those who have a diploma from elementary school works in what kinds of jobs in industry, or what problems what kinds of women in the population have. In order to reassemble or reaggregate his data, he is forced to resort to standard and arbitrary assumptions or to the results of other studies. Or he may have to fall back upon that which he considers true according to common sense.

For example, if a certain number of persons without diplomas live in the city, he might tend to regard this as an attribute of the oldest persons belonging to a particular ethnic group. Instead, it could be that a significant number of young people are in this condition. Obviously, this could be an important problem of that community to be taken into consideration. But the urban planner has few if any instruments that give him the capability of discovering it— educational administrators *may* have relevant data but they are ordinarily far removed from the urban planner.

Incidentally, the assumption that a certain number of old people are without diplomas—or even illiterate—should also be cause for some sort of intervention. That instead is generally taken as a normal fact when found or processed by the urban planner. One usually thinks only in standard modern terms of "assistance for the aged," that is, with physical survival and not with civilized living.

If the urban planner proceeds, then, along such lines of data collection and manipulation, he finds himself dealing with a series of fragmented images that are more imaginary than real. The initial reduction of the community to a series of segmented numerical entities no longer permits the planner to rebuild or otherwise approach persons in their totality. Nor can he any longer bring into view the presence of social groups and see their problems.

At present the urban planner is caught and held tightly in a trap by his own methodology. Total persons cannot be seen even with the most advanced technology in data processing, that is, with computers. Indeed, urban planners who use these instruments with their ordinary data do so only for the purpose of doing at high speed operations that would otherwise require a much longer time. Even if it is possible to use these instruments in interesting ways which tend to modify the very nature of urban planning, for the moment this has only been proposed by some and partially experimented with in most exceptional circumstances.[20]

Most use them, as I have said, to do traditional operations in a very short time. These are precisely operations involving abstraction and the standardization of the various elements in the community under study.

One can better understand what, in my opinion, constitutes one of the crucial procedural limits and errors of the modern urban planner by phrasing the matter somewhat differently. We might ask ourselves what model of man he has in mind and to what he intends to attribute the numerical measures pertaining to the community in which he has gathered the data. The answer is that it is the model of a man who is partialized or fragmented, whose recognized needs are the sums of certain separately determined necessities that are destined to be satisfied separately through specialized institutions.

One can, perhaps, object at this point that this process of reducing persons to numerical categories is the only possible way to achieve objectivity, neutrality, and scientific rigor. I have already underlined the point in preceding paragraphs that the very mechanism of the plan is already per se subjective and biased, especially in stratified societies. I would now like to add that in my view this way of "reading" the social fabric can give precisely the very opposite of the neutral results that a person making such an objection intends. In fact, if the planning process were such as to illuminate the problems, needs and interests of the complex of social groups in their holistic multidimensionality, at least some of the needs of the most deprived groups would become so evident as to be considered as having top priority in a way that does not happen today.

These methods, then, substitute the reading of real problems by standardized, stereotyped and abstract views of them. They give instead the opportunity to those groups that hold economic power and have a clear view of their own interests to pass these off as general public interests.

But let us return to the traditional procedure of the plan that we are analyzing. Besides the criticisms that I have formulated concerning the way the urban planner acquires his knowledge, there is another aspect of making the plan that is even more important, since it is even more essential for all of urban planning.

The current phase of capitalist-industrial society is characterized everywhere by processes of privatization which, among other consequences, result in solitude, boredom, neurosis, alienation, mental problems, criminality, delinquency and so on, all these being phenomena that are, for the most part, particularly evident in the city.

Well then, how does the urban planner position himself in regard to these problems? How does he think his plan should function with respect to these and other phenomena that, as we have noted in the last chapter, constitute some of the chilling ills of modern urban society?

Various answers are possible. First of all, urban planners can say that there are two separate problems. Some planners assert that these ills are "in" the city but that they are not "of" the city. Thus, the planner himself can perhaps be involved with them as a so-called private person, but they do not enter into his sphere of interest and action as a technician of urban planning. Alternatively, he can recognize that they are problems of the city but that they concern other specialized sectors than that of urban planning.

Neither of these two responses is correct, unless one defines urban planning in such narrow terms that not even the aims that are currently given as reasons for making a plan could be part of the discipline any longer. Furthermore, the negative effects of the organization of physical space on the problems mentioned above have appeared evident in numerous studies.[21]

The urban planner can react by saying that there are negative effects involved in the *application* of his plan; in other words, they originate from the wrong use of the plan. He, however, is not responsible for this. Such a response is obviously unsatisfactory and recalls a frequently held and similar attitude of other "experts" and "technicians" from other disciplines, among them the designers of nuclear weapons.

There are two other possible responses to the preceding question that can be treated together, although one is contrary to the other. The first is from the urban planner who says that he is very interested in these problems but holds that urban plans cannot really be effective in their solution. The other is from the planner who, instead, holds that these problems are within his competence and that he is presently dealing with them precisely through his work on making plans.

These replies must be seen together since, in fact, the errors that the urban planner commits and the situation of urban planning's instrumentation today are important for responding to both. On the one side, they explain to those who take the first of these two positions why in fact the urban planner is not truly effective. On the other, they show that if the urban planner has an impact on urban problems, it is an impact that exacerbates rather than solves problems. By accepting and strengthening the sectoralization and institutionalization that characterize the city, he helps to intensify urban problems.

We have already seen that the planning process is distinguished by its abstraction from the real man and from his class connotations. Also, in his model of urban social man, the urban planner submits or admits that the individual has a limited number of separate and separable needs, for each of which there exists in society one or more institutions created precisely to respond to them. This limitation and separation of needs on the part of the planner himself contributes to the fragmentation of the person, to the division of his life as a person and as a member of a social group.

It is important here to recall Marx's analysis of the relation between production and consumption. He said that "production creates the consumer" and, again, "the artistic product, and in the same way whatever other product, creates a public sensitive to art and capable of aesthetic enjoyment." [22] One sees, then, how the planner's typical model of man, distorting the product "plan," is transferred to the "consumer" of his product. Citizens subject to such plans become more sensitive to seeing man in these abstract, alienated and more and more alienating terms.

We can now establish a most fundamental point. All of the other errors of the urban planner that I have so far considered are without a doubt on another, lesser level. In the gathering or treatment of data, one can, in fact, overcome the inadequacies underlined earlier. One need only use more advanced methods and instruments which, even if they are still not commonly used in traditional urban planning, have been applied in some instances.

It is impossible, though, to modify the situation unless the urban planner assumes an entirely different way of introducing man in his totality and in his group and class reality into urban planning processes.

I have anticipated this conclusion by leaving aside the question "how" that I also ask myself and which I will treat in later chapters. I would like to continue now my examination of the traditional way of drawing up a plan.

Quantification and location of services and functions

In this first part the model of man held by the urban planner and urban planning has emerged. It is equally important to try now to answer another question: what model of the city, if any, does the urban planner have in drawing up his plan?

Once again, let us try to see it from the perspective of the operations he carries out. Once the famous data that we listed have been gathered, the urban planner considers himself ready to proceed to what we have earlier called the two fundamental operations: that of quantification and that of location.

The first quantification, once working estimates of the population have been obtained, usually concerns public services. The quantification of services is done by applying urban planning "standards." The goal of equity, that is, the desire to guarantee equality to all citizens who wish to use public, cultural and recreational services, is presumed. In some countries the standards, numerical parameters, are defined by a legislative measure. Once again, they are guaranteed by the authority of the state. (The general tendency in most countries is toward the legalization of such parameters.) These standards establish a certain area of so many square feet, yards, acres and so on for schools, green areas, parking lots, hospitals and other services or facilities that, on the average, citizens deserve.

But is it really true that such minimal standards are a useful and positive planning instrument? (We will treat another question in due course: is it true that such standards are necessary?) I think that an eloquent reply can come from just those new housing areas that were developed following the prescriptions of the standards. With few exceptions, living conditions in such areas are as inhuman and alienated as those of other areas where the standards were not applied. I am not speaking of their "adequacy" nor of how these standards were obtained. I do not think it matters if ten square yards of park per person are prescribed rather than eight, or eight rather than fourteen.

Beyond the numerical measures of such prescriptions, it is important to understand the logic that led various countries to introduce such standards as laws. They have been accepted or solicited by urban planners with the following type of argument for uniformity:

> The various laws and projects of urban planning tend to make the territory available to the urban planner and urban planning as a sort of "tabula rasa" on which they can arrange models of development (that is, designs) answering to diverse logics and conceptual approaches. . . . Only by introducing the concept of planning standards does one succeed in fixing criteria to which all plans will have to respond.[23]

Again urban planning reveals how it entrusts the securing of qualitative aspects through quantitative parameters: standards. The use of standards not only does not guarantee the attainment of urban quality but also testifies to the ever greater bureaucratic essence of planning. It is easier—for the officials—to impose common standards on citizens of heterogenous styles, forms and needs of life. Services seem to constitute the meeting point of two of the six ideological tendencies of urban planning identified by Leo Jakobson: the quantitative deterministic one, often called rationalistic, and the administrative efficiency-oriented one.[24]

In this sense standards represent the logical consequence of the principle that is at the base of all modern urban planning: zoning. Territory is subdivided into areas subjected to different regulations and in each area particular functions are allowed. This zoning was applied since the beginning of modern urban planning in order to rationalize and resolve problems that loomed up with the chaotic expansion of the new industrial cities of the nineteenth century.

One reads in books and dissertations on urban planning that zoning is the fruit of a rational vision that tends to see the city as a machine in which the gears are single functions. These are taken separately, with the appropriate regulations that the separate study of them permits one to promulgate or discover. With the appropriate special, differentiated treatment, they can reach good performance levels. This makes it possible for the city as a whole to function better.[25]

It seems possible and appropriate to substitute for such a model of zoning, born from a specialist perspective, another vision, another understanding.

Zoning has been diffused and applied as a criterion in much, if not all, current urban planning legislation, as the natural and beneficent transfer of the principle of the division of labor into the urban environment. Zoning radicalizes such a division by institutionalizing and bureaucratizing it as the division of life patterns in the urban milieu.

Standards are much the same. They also pretend to respond to human needs in the urban context, in fact, by institutionalizing and bureaucratizing these as the requirement of certain residential densities, of certain amounts of green spaces, also considered and evaluated separately.

One can, of course, point out that standards, which for the moment concern only services, serve to guarantee that areas are left in which these services can find a location. If there were no standards (or zoning), when a public institution planned to build a school or a hospital or an athletic field, it would no longer find available areas. In other words, standards serve to protect the population against speculation; to guarantee a greater degree of justice in the city.

Even at the risk of being considered blasphemous, I must assert that this does not seem to me to be the central problem of the speculative city. In fact, even if one cites the example of children who in urban areas are forced to play on the sidewalks, only two steps away from car traffic, I do not consider this the most unjust of the expropriations carried out against them and against citizens of every age. That, instead, comes from citizen impotence and alienation to which the procedures and forms of urban planning contribute far more than the injustices that have or can be alleviated by the use of standards.

In fact, the instrument of standards is inserted into the logic of the model of the city which modern urban planning has elaborated and carried forward. At the most, this logic proposed the rationalization of the topsy-turvy urban milieu of industrial society. It is clear, at least to some of us, that the problem can no longer be put in terms of rationalization, but only in terms of radical transformation of urban society.

Some scholars, for example, have begun to examine and to criticize the services of society. They also doubt or deny the presumed result of greater social justice. Public services, as organized and provided today, are an excellent means of repressing and directing society. This is so precisely because so many people depend on them, above all, those who have no private alternatives.

The use of urban planning standards is not, in my judgment, valid even outside the context of stratified capitalist society. That is, I do not agree at this point with those, and there are many, who accept its validity within socialist urban planning.[26] In fact, the image of a society and a city in which everyone has more or less the same desires, aspirations and needs can, as I see it, be identified today with the most developed capitalist societies. There the laws of consumerism can push persons to use product A rather than B or C, and nearly everyone has substantially the same anonymous model of life despite vast differences in wealth.

It seems to me that the way to an alternative society must lead to a challenging of capitalist society and of its planning standards. One can do this precisely by committing oneself to securing equal opportunities of expression for everyone, as well as the satisfaction of needs that, being human needs, are diverse although not for that reason unequal. If, however, one thinks that through standards one can achieve homogenization, one commits the error of believing that homogenization necessarily means equality in satisfying human needs. Thus there is still another error in the modal urban planning process: the assumption that it is right to use quantities that are fixed and uniform for everyone. The fact that uniform standards are specified for each function means the acceptance and confirmation of the heterogeneity and division of these functions, both among themselves and from other activities, and of the division of urban man's everyday life. Individualizing functions means homogenizing people.

To summarize, man's everyday life is composed of various activities that can be organized and carried out separately in space and time. As the division of labor and specialization progressively increases, the "division of life" in the city, which I have discussed at some length, is realized simultaneously. Urban planning has contributed to this process by rationalizing the organization of segmented physical space. The practice of zoning, in fact, is one of urban planning's major contributions to the division of labor and life.

The *Athens Charter* of Corbusier gave a distilled vision of these segmented space principles.[27] Although the charter is no longer in style and therefore is not frequently quoted, it continues to be reflected in the model of the city of most contemporary urban planners, as well as in today's forms of intervention for the solution of urban problems.

It is through an operation of zoning, then, that the urban planner concludes the preparation of his plan. When the maps of the plan, the body of the report and the norms for its realization as embodied in various codes are ready, the entire urban territory ends up divided into different land uses. One can make residences or an industrial area, a park or a hospital, according to the symbols that specify the appropriate land use of the various area zones.

What is wrong with this? Precisely the fact that zoning is the last link in the construction of a highly segmenting urban-planning instrument that does not address the basic problems of the divided and alienated society of today.

We have seen, then, that the fundamental principles on which the process of formation of a plan is based are the following:

· the reduction of persons and social groups to numerical, abstract and depersonalized entities;

- the standardization of their needs both in the sense of reducing these to a small number and in the sense of reducing them to numerical quantities;
- the separation of the life of persons according to "functional" parts;
- the obligation to carry out activities or to satisfy needs in different areas and in forms rigidly separated from each other.

If one adds a zone for offices, plus a few residential zones, some industrial areas, and, to this adds some parks and gardens, one is still quite far from producing a city. It is even more true that the sum of a number of hours spent in typing or in factory work, plus hours spent in eating and sleeping, plus recreation or relaxation in the park does not make a human life. Yet such arithmetic operations are actually what the planner does, given his models and modes of working.

This is the key also to understanding why the city and society are interpenetrated.

As I noted in the preceding chapter, man is alienated and dissociated not only by being dispossessed of his economic products but also of his cultural, political and social ones. And he is so dispossessed when his totality, his whole personhood, his integral personality is partialized by such treatment and he is put into such a segmented system. In the light of all that we have seen concerning the methods of making urban plans, we can conclude that plans are instruments that contribute to just such alienation of the individual.

As we have also seen before, plans function and affect people in a manner that is in inverse proportion to a person's status; that is, the lower his status, the more plans influence his life. Consequently, it follows that *given the way urban plans are traditionally produced, they are instruments of alienation and control that especially affect, and adversely so, the less well-to-do.*

The Plan and its Implementation

The foregoing conclusion might seem too dogmatic to some and simply unacceptable to others. One might object, for example, that I have focused all of my attention on the single character of the "urban planner" while, at most, only the outline of the plan is proposed by the urban planner. The real choices are made by the local government or by the organs it has set up for planning. One can also add that now the method of involving the population in the choices of the plan, that is, citizen participation, is more widespread, thereby further diluting the impact of the planner. Even granted that the urban planner commits the errors to which I have referred, this all means that the municipal administration first and then the others involved, the participant citizens, are in a position to make choices that can compensate for such errors. Thus, ultimately, the plan can be an instrument that maximizes *"the quality of human life in urban space and the equality of opportunity in using physical space,"* according to the theoretical presuppositions that I define as consensual objectives of urban planning.

Who better than the populace itself and its elected representatives can evaluate the actions and choices necessary to reach these objectives?

Indeed, I agree, but I do not accept the preceding objections, precisely because conditions do not now exist for these decisional possibilities, especially with regard to the population at large. The critical conclusions I drew in the preceding section are, if anything, weak. And the actual use of plans is without doubt even more negative with respect to the theoretical objectives of urban planning than the process of making the plan, which we just examined.

Once the plan is made, it becomes, as we will see, an "economic instrument" that is an important part of the mechanism of urban revenue. In that case too, the plan does not act in an egalitarian way, given the division of modern societies into classes.

Let us now take up the first point: who actually makes the choices involved in the plan? And then we will consider the immediately related point: what relation exists between what is established by the plan and what occurs in the urban reality with the plan's realization?

The choices in the plan

The traditional method, and in most cases one that is still followed today, is that when the urban planner has made some progress in the drawing up of his plan, he subjects his ideas and proposals to those who commissioned it. This generally means the urban planning commission or, at most, the city council or a committee thereof. For several reasons the type of discussion that then follows ordinarily results in few changes. First of all, discussion takes place on a level and with a "specialized'" language in which the urban planner is much more at ease than administrators or appointed or elected officials. Furthermore, the urban planner is generally considered more "competent" in the matter than the latter.

In other words, the administrator or official is victim of the trap that he himself and society set in this regard. In fact, in accordance with legislative prescriptions but also his own convictions, he believes that in order to resolve urban planning problems, the "expert" is necessary. Now, faced with the planner, the administrator or official can only accept his expert experience. The planner is paid for, among other things, having "experience" and for putting such "experience" at the service of the community.

My description may seem extreme. Actually, I agree that this reaction to the presentation of the plan does not always occur nor does it always seem to. But I wish to emphasize the point that today in most cases it is only *apparently* no longer so; in fact it is still the usual case.

Let us look closer at what happens in order to decide if we must substantially modify this judgment that, in effect, the choices in the plan are made by the technician, the expert, the urban planner. It occasionally does happen that the municipal administrators, or some of them, represent specific special interests to which a certain design of the plan must correspond. Thus, if the urban planner has not spontaneously succeeded in producing a plan to fit these interests, he must be made to, even at the cost of having to disrespect his "experience."

One thing that is important to emphasize, though, is that pressures or modifications that may be requested do not concern either the methods or the general principles that we have sketched before and according to which the urban planner works. Instead, they concern such matters as land uses anticipated or preferred or the location of services or facilities. Many times, in fact, the dispute is over the road layout and the resulting physical directions of the city's development.

It follows that the plan would be able to bring about some minor adjustments in favor of some of the more disadvantaged citizen groups if we suppose that the special interests that certain administrators or officials have do not concern groups of wealthy speculators but, rather, disadvantaged groups. This is not the case, however, and as we shall see, these small adjustments thus have little relevance to the general nonegalitarian structure of the plan and its nonegalitarian way of functioning.

In this kind of case, it is true that the choices are no longer really made by the urban planner, but by those who are in a position to influence him. Thus it is important to consider how often, when he does not make the decisions himself, he bows to the influence of the poor, the elderly or the most disadvantaged.

Let us take a look at the fact that the process of making the choices involved in the plan seems to be enormously enlarged in terms of numbers of participants in recent years. Indeed, by now local administrations and the urban planner himself solicit with ever greater frequency the population's "participation" in deciding on the choices of the plan.[28] But is the number of those who are actually involved in this decisional process and in the choices really enlarged? Or, instead, is this a method of "constructing" consensus over choices already made, as some scholars have defined the present participation process?[29]

It seems to me that these scholars are correct. What happens, in fact, in most current cases of so-called participation? The process of outlining the plan remains substantially unaltered. At a given point in its elaboration, the proposals are presented to the citizens, within the context of so-called participation, as well as to the administration. Once more, the element of expertise, about which we spoke earlier, enters into play. The expert (in this case the experts, because administrators also participate as "experts") chooses the ground for discussion. Once again, that ground is his specialization. When he presents the population with his plan, this instrument that renders reality abstract, the plan's correlations with reality are, at best comprehensible to perhaps most of the experts and very few nonexpert citizens.

An example will clarify the situation. Let us think of a very beautiful design of a car in axonometric perspective. Thanks to representational techniques, everything seems understandable. The distribution system of electrical energy is in one color, the fuel system in another.

Now, by presenting this design to experts, perhaps—I emphasize "perhaps"—they would be able to judge the characteristics of the car. Probably, even if the design were very accurate, they would need other information and, above all, would need to experiment with the car while it was running.

Certainly, those not "experts" could not even begin to pass judgments on the basis of that representation of the car. By trying the actual car, however, they could at least evaluate its adequacy in relation to their own needs and desires, even if they were unable to judge it in relation to other needs.

Well then, what can one say when facing the maps of a plan? Few can be anything but dependent on the interpretation of it given by the expert. Currently, then, citizen participation is usually nothing other than the expression of an opinion that can only be manipulated and dependent unless it happens to be an expression of outrage against a plan's provision that is injurious—and often that is manipulated as well.

It is useful to note once more how the conditioning process, the product's effect on those who consume it, works. Although I have not previously entered into the matter of who are the "consumers" of the "plan" as product, I would now like briefly to consider it. The major direct consumers are the private developers and owners and public authorities.

A very popular theme of planners is that of strengthening the public use of planning. It is often advanced in proposals meant to give urban planning a capability of functioning as a reforming instrument of society. In other words, if plans are not to remain tools for exploitation, for speculative developments serving narrow private interests and for the few at the expense of the many, many reformers say we must ensure that among the consumers of the plans there are a greater number of public agencies and institutions and ensure that private interests are restricted and controlled as to where and when they can engage in using land. This can be summarized with the famous phrase, often repeated, of "increasing the public presence" in the management of plans for better land development and redevelopment.

But what will happen if public institutions do become the principal "consumers" of the plan? It has occurred, not only in various so-called socialist societies but also in capitalist Italy where public and semipublic corporations abound. The public bodies are just as conditioned by urban planning to see and operate a fragmented, sectoralized city. Some of the worst private speculative developments are now impossible, but with public agencies as consumers there is even greater likelihood of what Mumford terms the Invisible Machine coming to more effective, efficient and speedy fruition.

We stress that Mumford was not speaking, more than a decade ago, primarily of the Soviet Union nor even of an increasingly planned American society, but of modern advanced industrial societies regardless of the form and techniques of government:

The many genuine improvements that science and technics have introduced into every aspect of existence have been so notable that it is perhaps natural that its grateful beneficiaries should have overlooked the ominous social context in which these changes have taken place, as well as the heavy price we have already paid for them, and the still more forbidding price that is in prospect. Until the last generation it was possible to think of the various components of technology as additive. This meant that each new mechanical invention, each new scientific discovery, each new application to engineering, agriculture, or medicine, could be judged separately on its own performance, estimated eventually in terms of the human good accomplished, and diminished or eliminated if it did not in fact promote human welfare.

This belief has now proved an illusion. Though each new invention or discovery may respond to some general human need, or even awaken a fresh human potentiality, it immediately becomes part of an articulated totalitarian system that, on its own premises, has turned the machine into a god whose power must be increased, whose prosperity is essential to all existence, and whose operations, however irrational or compulsive, cannot be challenged, still less modified.[30]

Our point here is that shifts toward public ownership or public planning or even away from public ownership or public planning are irrelevant except insofar as they may prove to be distracting, time- and energy-consuming efforts that miss the underlying and increasing human problems of our age. Or they may even continue under public as under private aegis to contribute to molding the partializing, functionalizing mind-set and behaviors that are necessary for the hierarchically organized institutions of specialized and sectoralized modern society.

We must not forget also that consumption influences production; that consumers also influence producers in this circular process of influence. A basic quality of urban planning as it exists, and as it existed earlier, is just that fact that society is divided into a myriad of institutions; it is not complex but it is made complex and fragmented. From this viewpoint, then, the plan is essentially shaped by inputs, demands, coming from such separated, specialized institutions.

We return in this context to the matter of citizen participation. Its character today makes it far from being a democratic opening of the decisional processes. On the contrary, it occurs while the plan strengthens the capability of organized institutions to make the number of "consumers" of plans ever larger. Consequently, counter to the major thrust of citizen participation, institutions spread the divided and segmented models of man and society, models that the plan reinforces, even more widely.

Perhaps one would disagree with the image of citizens as passive recipients or consumers rather than active demand-makers. Those who participate, however, usually make demands or proposals that are marginal and secondary—or are treated as such—if they are outside the partialized institutional logic of the plan.

In evaluating a plan, among "citizen" participants there are often experts in various matters subject to the urban planning; that is the situation even when a plan is to be examined by so-called ordinary citizens. There may also be some citizen participants interested in following the unfolding of the plan precisely because they have personal interests to protect. In some such cases, it is quite possible that these persons or groups do succeed in organizing a consensus against the urban planner and the local officials involved. This is the usual form with which special interests have succeeded in affecting planning processes.

Indeed, these not always unqualifiedly positive varieties of citizen participation are not noted here to suggest that planning choices should be restricted to the smallest possible number of citizens—to the obviously responsible, well-intentioned and public-spirited citizens—in order to protect the plan from undesirable consequences of speculative pressures. Rather, they are here to help in understanding that it is necessary to face the real problem: that is, how to modify urban-planning processes in order to have real participation rather than the present too often directed participation or pseudoparticipation. In the chapters following I will discuss this vital problem.

I return to the question about who makes the choices in the plan. In order to answer, it will be useful to keep in mind what has been said so far. When the urban planner is charged with making a plan and begins on it, he is not entirely free. He must follow and respect a series of laws specifically concerning the practice of urban planning. These are in addition to other relevant statutes. But what are these laws if not the manifestations of agreements or compromises of the interests of power groups that have produced them? What are they if not concrete expressions of the necessity to control both the needs and the development of society, formulated with the support of the knowledge of urban planning experts? It is the latter who have articulated the forms that those interests and needs should take.

I emphasize these points to stress that the basic urban planning legislation, made with the collaboration of planning experts, gives to other planning experts criteria that conform to the preparation, the professional baggage and the methods of urban planning. Essentially, the inputs or conditioning effects in plan-making that derive from urban-planning laws are the demands, specifications and limitations specifically suited to the needs of the urban planner.

Why is this important? Because it permits one to conclude that the legislative foundations of the plan are shaped by the urban planner according to the concepts that he uses as a technician, a specialist, an expert.

Speculation and special interests exist, but they are hidden inside the "technical" and seemingly impersonal standards and procedures required by law. The seemingly general and nonmanipulative calculations of standards do have in fact a great deal of elasticity. Part of the reason is the aforementioned necessity that the plan have margins of flexibility in order to vary with circumstances, but this elasticity is also due to the fact that many interests are not adversely affected by, or at least have adapted to the principle of, zoning. They have become skilled in influencing the choices of locating activities and possible land uses.

Thus, for example, often no one will object that certain areas are reserved for educational use only and others for athletic use only. On the contrary. It also suits the privileged and is in their interest that these needs be satisfied by public actions. If, however, someone has specific interests in such areas, he may try to prevent such facilities from ending up on his land.

To return then to our answer about who makes the plan's choices, we can conclude that the plan is basically shaped by the principles, modalities and decisions of the expert, the urban planner. Within this frame the existing partial interests of the community affect choices. It can happen that these interests are so strong and broad as to be incompatible with the plan's prefixed grid. In such cases there are few possibilities; either the plan is adjusted with respect to those interests or the interests are forced to conform to the possibilities foreseen by the plan. Such choices are usually made by the planning expert and the expert administrators. And those principally involved in the process of choice-making tend to be submissive to powerful interests. That is, the urban planner and the local administrators are generally predisposed to compromise with the so-called vested interests, usually real-estate or industrial developers and their financiers.

Someone might ask what would happen if both the urban planner and the administrators wholeheartedly intended to resist private interests and to carry forward the asserted general interests of the plan and those of the public. In other words, what if they should try to reach those two objectives of quality and equality in the use of urban space that should be at the foundation of urban planning? My view is that they will most likely fail if they intend to use today's urban planning instruments in their present form and traditional manner. One of my goals is precisely that of clarifying what one should begin to change in order to increase the margin for success in such efforts.

We have already seen not only the technical but also the political distortions that mark the making of the plan. But does the possibility remain that the management of the plan may succeed in modifying this sad state of affairs if it is a plan substantially accepted in its nature both by administrators and eventually by those who participate in its workings?

In order to assess this and, therefore, also to see whether or not my pessimism is well founded, it remains for us to examine what happens after a plan has become law. In other words, once the municipal administration has adopted a plan, what happens to it after it has gone through the necessary bureaucratic routine and has been approved by the competent authorities?

How an urban plan functions

A number of aspects must be considered before one can evaluate what a plan really produces. For the sake of clarity, I think it is useful to establish what a plan consists of when it is finally made. Although there are slight differences from one nation to another, we can say that the material product in all is substantially divisible into three parts: the basic report, the maps, and the norms for realizing the changes the plan calls for. These must, of course, conform to or respect various rules or codes concerning sanitary and safety aspects of present and future development.

The report is the written part. It describes and justifies the choices of the plan. It is the part that contains the elaborations of the data and presents information on the forecasted growth or decline of the population on which the plan rests and contains historical information on urban settlement.

The maps of the plan generally reflect the current state of affairs (present land uses) and the so-called project or the projections, plans and hopes for future land uses.

The norms or rules for realizing the plan, finally, make up the modalities to be respected for each zone when one begins to build or locate the facilities permitted by the plan's provisions.

For convenience's sake I will refer once again to an automobile, using an analogy that is not perfect but does suggest some important and useful points for reflection. The maps of the plan constitute the automobile itself; the norms, the instructions for its use; the report, the description that the producing company usually gives of the automobile itself.

When the metal object called an automobile leaves the production line, it can become, it has the potential of being, through the use of one or more human beings, an actual automobile. It can also, however, lie unsold and be destroyed without ever having been used, without ever having become in this sense an actual automobile.

One can say the same for the plan: that is, the plan is realized only in the moments in which it begins to be used.[31] Further, the realization of its essence as a plan is not a constant but a process.

In its totality the plan points to a configuration that the use of physical space in a given city should have in a given period in the future. It is per se a static configuration but it indicates the transformation between a present and a future situation. It should, thus, imply a series of transforming actions rather than identify a configuration that will remain static once it is reached. It should open up further necessary and dynamic processes.

Now, granted that a community possesses a plan and wants to use it—that is, it wishes to obtain the configuration that "its" plan foresees—is it in a condition to make that happen? Returning to the example of the automobile, this is like asking: granted that a person has bought one and wishes to use it, is he in a condition to do so? The automobile answer is obvious: yes, if, for example, he knows how to drive.

If he does not, he has two alternatives: he can either learn how to drive or he can have someone who knows how to drive do it for him. In the latter case, he must hope that this someone will fulfill his desires or will understand his needs and drive him as and where he wants to go. I do not mean to be facetious when I say the answer for the plan is analogous: those who know how to do so can use the plan.

We must project this into the reality of a stratified society. Im-

mediately one can see that various situations exist, two of which are
especially interesting. First there are those who—as we have already
seen—own or can easily buy this knowledge of how to use a plan;
these are generally the wealthy or powerful. Then there are others
who know nothing or think that there is no other alternative but to
trust someone to use the plan well, that is, in their interest. For the
bulk of citizens this someone obviously can only be the local officials.

For the moment we assume that all the basic inadequacies and
distortions that we have revealed with respect to the objectives of
quality in the use of urban space have been left out of the plan's
composition. Let us see if, even then, the local officials are really in a
condition to make good use of the plan.

Were the user of the plan truly analogous to the buyer of the
automobile, the problem would be merely that of learning to "drive"
the plan. The aspiring motorist learns to drive his car and often at
least learns about the essential functioning of the motor, reserving
the right to go to the mechanic when it is functioning poorly. The
plan, however, is not "driven" quite this way. A substantial difference invalidates the analogy: the motorist can learn once and for all a
series of rules that assure him that his automobile will function in
such a way as to satisfy his purposes, even if they do not guarantee
that he will become a good driver or have no accidents. This is possible because he has in hand a so-called deterministic system, one in
which all the parts can only function in a coordinated given way (or
not function in other ways).

Local officials, however, must administer a city, which is a so-
called probabilistic system in which no relation is preestablished in a
deterministic way. As we have seen earlier, the plan itself represents a
probable, a hoped-for and a presumably optimal state of the system
in a given future period. Such a state or set of conditions has been
envisaged in the plan as the anticipated result of a series of anticipated actions. Such actions are to be performed not only by the administration, but also by other organizations, public and private, and
individual private citizens operating in the city.

Such actions can occur in a certain temporal sequence and produce certain anticipated effects; but they can also occur with other
rhythms and produce quite other effects. Or they may not occur at
all.

Let us look carefully now at the situation the local government
finds itself in once it has the plan in hand with regard to anticipated

actions and their timing. The master plan, for example, may foresee the realization of various residential housing developments. But if private landowners do not actually build in areas zoned for such residential uses, what can the local administration do? Very little, if anything. This holds true even if one is dealing with a public agency as owner rather than a private property owner. The presence of a certain forecast in the plan does not, save in exceptional cases, call for obligatory actions.

The same municipal administration often cannot intervene according to forecasts of the plan because of adverse financial circumstances or declining municipal revenues. But can it perhaps influence the timing of actions foreseen by the plan? In fact, usually all changes projected by the plan, except for a few specified matters of timing, are potentially realizable simultaneously. Thus, for example, very little can be done usually to prevent a given part of the city from urbanizing before the time forecast in the plan, thereby delivering a major blow to the entire urban organization anticipated in the plan. And as I have emphasized, social classes and groups that have the means to act rapidly and independently enjoy a high degree of freedom simply by conditioning, reducing and influencing successive actions. Those who, in contrast, must depend on public action know, in general, that they must wait and wait. And at the end they will be dealt with from the more limited possibilities still open.

In the shaping of a specific future state to which the development of the city should conform, the plan, then, has these deficiencies. First of all, the agents who are to bring about the passage from a present to a future state are not identified, except for the rough distinction of public or private. Moreover, the passage is ordinarily to be effected within the context of substantial or predominant private ownership. Nor does the plan take into account that the processes that it presupposes are probabilistic nor that some, or all, of the actions proposed can have unforeseen consequences, especially but not only when free private actors are involved.

Various proposals for modifying the plan's procedures by introducing short-term programs are only partially satisfying. They do not correct such basic deficiencies; they can only act as checks or eventual corrections in a shorter period of time and that is not the central problem.

In realizing the plan, there is also another problem that is undoubtedly greater than all the others noted thus far: *the plan has no explicit criterion for evaluating the quality of its anticipated actions, public or private*. It can only judge in quantitative terms related to zoning. Therefore, in no instance can the plan be used to guarantee the control of the "quality" of the practices or actions through which it is put into effect.

Such quality control is impossible in the management of urban plans. By "quality" I mean social quality. This includes aesthetic quality which of course is part of social quality, but for the most part I am referring to the quality of social life as it is linked to the use of urban space.

It is difficult to elaborate criteria in this respect because, as I have emphasized, we are taught to evaluate physical and social systems separately and differently, to the point of not even having a language adequate for expressing the categories, properties and quality of sociophysical space.

What I want to stress here, though, is that neither urban planners nor local officials feel, or have felt, the lack of these criteria. If this were not so, in fact, they would have sought different instruments instead of continuing to produce the same type of plan. Perhaps some vaguely feel the lack without being able to say exactly what it is or how it is to be met.

In the absence of criteria for evaluating urban sociophysical quality, there are surrogates that are used in the making and, in other forms, in the realization of plans. These criteria, however, refer to the meeting of economic needs, of functionality and of efficiency — relative to the urban scene viewed institutionally and compartmentally.

What do I mean by this? Let me begin by defining what I have called the criterion of "the meeting of economic needs" of the plan and its realization. In most of the experiences of urban development since World War II, the urban planner and local governments have demonstrated on innumerable occasions that they believe social objectives of urban quality and equality have their equivalents, "surrogates," in the "good economic use of physical space." In other words, it seemed that the well-being of the city was predominantly if not exclusively understood as, and measurable in terms of, economic expansion.

Since in modern times economic expansion is usually identified with industrial and commercial expansion, generally every plan in its forecasts and every local administration in its use of the plan have tended to favor industrial and commercial development and their immediate or indirect needs. Apart from industrial developments that occurred before plans were generally operative, I am referring here to all the forecasts made for industrial and commercial uses with little consideration for such other uses as agriculture.

Someone may remark that these are the rules of modern society. Even on the basis of Marxist analysis of urban problems, it might be asserted that it could not have been nor can it be otherwise, given the fact of the occurrence of industrial development. The latter has been the conditioning and shaping element of societies and the modern city as well.

To some extent I share this conclusion. At the same time, I find it to be the ground for stressing both to Marxist urban planners and administrators in such countries as Italy and to non-Marxist urbanists elsewhere that one cannot hope to have a great degree of success in creating a better city or a better society through planning by yielding to or accepting the logic that industrial expansion is the prime necessity for urban health. That logic depends on a simple, apparently reasonable and appealing but erroneous assumption of inevitability.

One might assert that in the past there were no real or acceptable alternatives, that the only alternative was to have no industries, to have no jobs. My reply would be that a different kind of industry and development was possible. And certainly, one can have, and one must seek to have, a different kind today.[32]

From the beginning of the industrial epoch one could have aimed at another kind of industry, industry that might have been better integrated with other life sectors. There could have been less destruction of the activities of artisans, for example, or of the artisan element within industry. It seems valid to say that too much, indeed, nearly exclusive, attention and uncritical acceptance has been given to industry, and the forms it has chosen to take, by urban planners and administrations of every ideological or political persuasion. Obviously, they could not have resolved the problems by themselves, but surely they could have contributed to a solution in a crucial way if they had not had such a passive, hands-off policy toward industry's growth.

The tendency of national politics in the United States and in Italy has been to facilitate and favor certain kinds of industries and industrial development. Those people who have economic and industrial power, as I have already emphasized, are not afraid of the provisions of a single plan. This, however, does not justify the fact that in the great majority of cases municipal administrations have not seriously questioned the kind of industry whose installation and expansion it was accepting or facilitating, except in an occasional instance of a heavily polluting industry.

Nor, in the great majority of cases, have urban planners seriously asked about significance or implications for the general human conditions of the given population of the size of the industrial areas or of the size or character of shopping centers scattered throughout their plans.

All of this might seem to be a useless recrimination of a past that can no longer be changed. This is not so. Once one is convinced that something more and something different could have been done, a long step forward has been taken in understanding that the same is true of the present relative to the future. In fact, in the current phase of advanced industrial society or of the society now being born, two perspectives can be clearly delineated. The first tends toward a perpetual technocratic and technocratically directed type of development. The second, instead, is aiming at an ever more human and small-scale organization by learning from the distortions that so-called development has brought with it.

It may seem that present society, with its artificially created and growing complexity, requires increasingly more complex and sophisticated mechanisms of centralized programming and control. It is precisely such new technology, however, with its increasingly more advanced methods of "computerization" and "miniaturization" that can increasingly substitute for hardware the more flexible systems of software. As a consequence, industrial and commercial organizations can also be projected on an increasingly small scale and be controlled by nonexperts just as feasibly as by "superexperts."

But the taking of one or the other of these two routes is not simply a matter of free choice. It is clear that the interest of dominant groups lies in the first perspective because it facilitates their maintaining control.

Are there any ways in which the second perspective can be facilitated? How might it proceed? Many writers such as E. F. Schumacher and David Vail have taken up these themes in their work.[33] They suggest that through the use of new technologies or a new use of traditional ones, decentralization and small-scale production are indeed possible. New informational systems are capable of facilitating self-management.

We stress in these connections that such innovations can begin and develop only with the opening up of institutions. These new opened systems must have as their objectives not functionality and the traditional economic emphasis but valued human experiences on the part of whole people. Participation of all those persons until now marginal and far from the decisional processes must be permitted.

It is possible, then, that the process of planning of the city could be an important occasion for administrators and urban planners who want to open up an institutional process and guarantee to urban people a chance to participate in the decisions that are basic to their everyday life.

The opening of the planning process means more than opening up only a part of it, that is, the management phase of the plan when it is already made. It implies opening the *entire* decision-making process beginning with the gathering of the first data and continuing throughout the period of application of the plan. The moment of decision is not only when choices are made on the basis of the plan's maps or norms. *Decisions are processes. People cannot participate in any real sense if they are left out or kept out of the process, called only when a "yes" or "no" is to be given and then sent home, leaving others to continue the decisional process.*

Naturally, it seems more efficient if only a few make the decisions. The exclusive type of organization that the urban planner accepts today from society and that he carries over into the master plan also seems more efficient and functional. But efficiency and functionality, much as "industrial growth," are sacred cows of contemporary society.

It is on the basis of such values that the urban planner organizes one area for a school, another for an athletic field, another for a hospital. Each is entrusted to a separate institution, most likely centralized and large, in order to be more functional. Such specialization is the touchstone of urban society as of urban planning.

Some may object that there is nothing wrong with this; that, on the contrary, the fact that the public administration seeks to regulate and does regulate its own actions according to criteria of efficiency and functionality is healthy. Furthermore, it may be asserted, this actually permits all citizens to be treated uniformly and fairly and standardizes actions separated from each other by long intervals of time. This, I respond, is a myth.

Why are these criteria so vaunted, but only selectively applied? In fact, rarely do those on the public side speak a word about the cost-benefit ratio (efficiency) in the case of industrial development. The latter is simply good. It is taken for granted that the public administration's expenditures for industrial development are worth the results, assuming such development occurs. This is also the case when expensive infrastructures are constructed that are "functional" only because they guarantee the functionality and efficiency of, and thus profits from, private investments. Rather than assuming such expenditures benefit the public, they are instead ideal occasions really to try to apply the criteria of functionality and efficiency.

There is a second reason these criteria are myths. Even if they were to be applied by public administrators in choosing what to do or not to do, these criteria are not now and would not subsequently be respected in the workings of the institutions, paralyzed as they are by the chronic ills of bureaucracy and characterized by bureaucratic inefficiencies.

Urban society has developed further the divisions of functions, institutionalization and specialization precisely in the name of rationality, functionality and efficiency. It has thus had an increasing degree of injustice and inhumanity.

The citizen who is not especially privileged is burdened with having to resort to various specialized institutions. He must somehow thread his way through a maze of vast complexity even if his question concerns only the single matter of his health. He does not even usually receive in exchange for the complex specialization a more efficient, a more rapid or a better qualified or functioning service.

In all this the master plan has played the role of rationalizing the mechanism of development of modern society; legitimizing the commercialization of the use of urban land and land speculation and containing the demands of the less well to do at a tolerable level while reducing the possibilities of conflict in the city by providing for some of their needs for services especially, and hindering the city's transformation by serving as a containing and conservative force.

These conclusions might be erroneously interpreted and provoke an immediate unjustified inference. If an urban planner, as I am, says that currently plans are useless and damaging, then a critic may assume the planner has said it would be better to throw the plans away and to stop planning; it would be better to leave the city in the hands of "free enterprise." This, however, is not my position.

I believe that planning is necessary. It is an instinctive attitude of man that has, especially in the modern epoch, gradually assumed unnecessarily specialized and institutionalized forms. The negative aspect, in other words, is not constituted by planning per se. By institutionalizing this process and making it a specialization, however, the present situation has come about in which the right to plan for the many is reserved for a few. This is true not only of the urban planner but of every other kind of professional planner.

Particularly in some countries, as in the United States, urban planning has in recent times demonstrated how monstrous it really is in practice. I recall episodes of urban renewal that have occurred throughout North America that realized large urban infrastructures or renewed areas useful and important to some members of the community but for which others paid and paid heavily. These others were the most disadvantaged. But this precisely is the reason for supporting the need for planning rather than a laissez-faire system.

Society, especially the one in which we live today, is not an angelic ensemble of human beings, each ready to acknowledge the human needs of others. To obtain social justice in the city, or better, to reach those objectives that I have frequently summarized as the attempt to create a better quality of urban sociophysical space and a more equitable access to such quality, we do need plans and planning. My point is that for these objectives the plans of today do not serve; they are fundamentally flawed. Our task, then, is not to use them while trying to improve them, because we need basic innovations rather than adjustments or reforms.

I referred in harsh terms to urban planning as monstrous. Sometimes the dislocation, physical, social, psychological, of groups of the least fortunate citizens is a visible, obvious brutality. Most of the time, however, the brutality is more covert, yet as fundamental. It is a process of putting or keeping people out or controlling them through a humanly degrading and reducing process of doing unto and for others what others can better do for themselves—working with each other. It is one of the myriad modern forms of rule without representation.

Someone might object that my analysis does not take into account the current process of critical revision, that is, the new proposals that have been made in these last years. In other words, among urban planners themselves the scarcity of results or the distorted results produced by planning have generated a process of criticism and self-criticism. That process seems to have produced some concrete proposals that, in their turn, seem to be capable of overcoming what has been called by some a transition period and by others a crisis period of the urban-planning discipline. Let us see what the new proposals are and what, if any, are their innovative aspects.

Urban Planning and the Transformation of Society

In the field of urban planning there has been a significant succession of experiences over at least several generations. The scarce results obtained and the at times dramatic events provoked by the conditions of urban life have obliged many planners, particularly in these last ten years, to reflect on their discipline and its tools.

In this section I intend to argue that the mountain of articles, reports and books dedicated to such reflection is characterized by a scarcity of innovative proposals and also by an absence of suggestions of new "paradigms" or models.

One can agree with some writers that never as in these last decades have approaches to planning become so multiple and differentiated. I have already cited two of the six ideological perspectives proposed by Jakobsen, and his is only one of many such categorizations. It is, however, banal at best to consider this multiplicity of approaches a mark of a democratic society, as one tends to do especially in North America. Pluralism in planning approaches, just like political pluralism, is often regarded most positively: as an expression of diversity of thought, which diversity must be safeguarded and maintained.[34] When we begin to examine the new positions and proposals, we realize that in fact they are substantially alike from country to country. One crucial identical characteristic they share is that they constitute the expression of dominant power groups. As such, they are more or less consciously oriented toward increasing the margins of the power of those groups rather than toward resolving urban problems. Thus, this pluralism is a pseudopluralism.

Faced with the failure or the trivial relevance of urban planning, there are those who seek explanations by pointing to the insufficient use or application of plans. In other words, those who hold this position maintain that the potentialities of plans are not realized and/or that these plans have not been sufficiently applied. As a result, the research and proposals of these apologists seek to "perfect" the technical instruments of planning in order to put planners and politicians in a better position to foresee and influence urban modifications.

This position is much more widespread than one may think. It is closely bound to the maintenance of the status quo, to the maintenance of the current forms of power and its organization, including the position's proponents' own places and roles therein.

This position extends over many in the large range of those who are involved in constructing urban models. The urban model-builders, many using computer-simulation techniques to probe how the model and the city will operate in future, stress the rationality of focusing clearly on a few explicit dimensions, however oversimplified, rather than vaguely on many, often difficult to specify dimensions.

The position also takes in those who propose to substitute for traditional urban-planning processes, based on master plans, so-called processual planning. The apparently radical transformation proposed by the latter lies, in fact, in setting much shorter deadlines within which forecasts, and consequently their verification, can be made. "Technical plans" thereby give way in the eyes of some to "operative plans."[35] According to these advocates, this procedure produces a fundamental impact on the very content of plans, which become less utopian and more rational and thus more suitable to the actual conditions of the urban environment in which the plan functions. Thus this procedure initially makes its impact on the methods by which a plan is formed and then on the approval and management of plans.[36] Our major criticism of such a proposal for change will be made shortly.

Among those who want more planning but do not really evaluate its nature, it seems that one should include those who see the need to connect urban planning to national economic planning in order to make the former effective. In the United States for example, the idea that economic planning at the national level is necessary has gradually become widespread although its popularity fluctuates with changes in the political economy and in Washington. The model is the indicative plan that has also been adopted in France and Japan. Instead of merely vague goals, more concrete targets are given so that a plan's actual accomplishments or failures can be noted and analyzed and done so more frequently over a given period of time.

Martin Meyerson, of some note in American urban planning in the past, remarks that among the advocates of this approach, and he himself is one, are promoters of the now inactive Initiative Committee for National Economic Planning, a group with a broad political composition.[37] Meyerson, addressing urban planners without going into a deep examination of the reasons for so large a consensus, says that the "next challenge for urban planners is in linking local and national economic planning."

Why? The major reason seems to be that in a time like the present, characterized above all by increasing and uneven pressures on scarce resources as well as increasingly complex national and international economies, planning becomes a vital element. Urban planning without national planning in a national economy cannot be effective and national planning, planning at the center and top, presumably needs planning at the bottom and base.

Another group, one that does not apparently share this perspective, also belongs with those who intend to "strengthen" rather than basically change planning. This group criticizes the excessive homogeneity of urban planning and argues for different types of models based on particular and varying dimensions of cities. However, what they have succeeded in producing, in fact, are only such things as superficial although correct criticism of the methods used in the planning of small towns, to the effect that those methods are a miniature reproduction of those used for large urban areas, and appeals for study of "appropriate small-town planning techniques (the use of the plural 'techniques' here reflects the diversity among small towns)."[38]

Here, too, we have an example of criticism that ends with efforts or requests to elaborate more perfect technical instruments, even when it seems initially to be directed against the forms of abstraction and standardization integral to traditional urban-planning methods. It is these methods and the models upon which the methods are based that need *correction*, not elaboration that leads to slightly modified methods with more custom-tailoring and a less standard application according to where they are to be used.

Another proposed planning reform must also be mentioned. For a number of years advocates of city-county governmental consolidations envisaged more effective and widespread use of urban planning throughout the single, enlarged jurisdiction. Currently, proposals for ever larger metropolitan "regional" consolidations or multistate intergovernmental districts are being pushed. These are, similarly, to be the forms for expanding urban planning to the territorial and organizational magnitudes at which, presumably, it will experience a quantum leap of effectiveness.[39]

Thus the apparently wide spectrum of positions and proposals for planning reforms leads back to a single attitude: that of assuming or reconfirming the validity of the esoteric and professionalistic forms that planning has slowly taken on since the last century. In other words, it has become a discipline based on an inheritance of techniques and methodologies that have never been called into basic question. As we have seen, the reforms are merely proposals for the increased use of this inheritance, albeit with small modifications.

The models of man and the city that ground traditional urban planning are not submitted to critical examination. Above all, the role and the right of the planner to interpret the community and to plan its future is confirmed, and that is also the case with so-called processual planning.

It is interesting to note that one expression of this acceptance of the role of the planner, which I define as conservative, is curiously enough also formulated by planners who are ideologically more to the liberal and left side of the political spectrum. I am now referring to those planners who, in looking at the past history of urban planning, criticize it both for having gone astray and for being unrealistic, "unrealistic" because it did not assertedly take into consideration the specific overall political situation. Urban planners pretended that urban planning was succeeding in effecting social change, although it was only able to do that, if at all, under very different conditions of power-structuring. By becoming activists in efforts at social change, planners have also abandoned the disciplinary terrain and, it is asserted, have gone astray. That is, their task no longer included continued research connected to those specific sectors associated with planning methodologies and techniques.

This is substantially a criticism of the urban planners' supposed departure from their disciplinary field to become amateur (and unsuccessful) politicians. Interest in playing such political roles was frequently expressed both in the United States and in Europe by professional planners who during the sixties became advocates of the opening-up of decisional processes, of grass-roots planning initiatives and of citizen participation.

Critics of such so-called politicization of urban planning have more recently substituted a vision of an urban planning politically committed but in purely professional terms. This is not a paradox because the commitment depends and must depend only on the conscience or ideology of the urban planner, whether he puts his professional urban planning skills at the service of progressive and democratic administrations; whether he uses his professional instruments to interpret the needs of the urban poor rather than the more well to do. In other words, the politicization of urban-planning operations, accepted or even desired by these critics, is not regarded as intrinsic to a "value-free" urban planning. The politicization of planning is not something integral to the nature of planning. It comes and must come as an extraprofessional commitment by those making a plan. It must come from outside the discipline, from outside the purely professional, the purely technical, the neutral techniques. It must come not from how urban planning, its tools and its techniques are used, but for whom.

This, I believe, is one of the most widely shared perspectives at the end of the seventies among urban planners who define themselves as advocates of the ideas of social justice and renewal. Such is the position inspiring many urban planners of the left now active in Italy. For example, they have worked with such local administrations as that of Bologna, which has for some time now been managed by the Communist Party, and for Socialist and Communist coalition administrations in other cities. This position, for the most part, is also the one of "liberal" reformers and the smaller number of more radical planners in the United States.

If we examine a recent experience in Cleveland, we will better understand this position and why it is, despite appearances, still a technocratic and bureaucratic position and why it fails to touch those points that are essential for the renewal of urban planning. Let us see why what Herbert Gans hopes for in regard to Cleveland is not tenable: "the Cleveland report may signify a radical change in American [urban planning] thought and practice."[40]

The urban planners involved in the Cleveland experience did not begin with the pretext that planning operations could be used in a sociopolitically neutral manner or with neutral effects. They made it their considered objective to use their instruments and professional expertise in favor of the poor.

In Davidoff's words of approval, as planners they participated in a process of "redistributive justice" by planning "for the poor and not for the affluent."[41] He continues:

> They assume that the greatest problem in the city and the one to which they may make a contribution is that of unequal distribution of material wealth, of poverty of whole groups and classes, of differential access to services and facilities and, finally, a situation of poverty in a situation of static or even declining resources.

We reject those goals, despite their appealing character, because in our view that is a misstatement of the nature of the primary urban problem. The major problem, the core problem, the heart of the problem, is that the less well-to-do social classes are continuously subjected to the greatest expropriation of decisional power in the urban environment—although other classes are as well.

Thus, it seems a great deal more important that planners help people gain more decisional potency, if possible by participating with planners in planning, than for planners to help people gain more money or services by planning for those people. Nor does the goal of decisional potency mean "promoting a wider range of choices for those Cleveland residents who have few, if any, choices," in the words of the Cleveland planners.[42]

What, precisely, is the kind of planning done now in Cleveland? In the sixties the notion of advocacy planning was introduced with the assumption that the impoverished and the inarticulate needed a professional planner to help them, to work for them, to represent and articulate and advocate their interests in a political game where the official urban planners had been on the side of the "haves," the articulate, the affluent. Of course the notion of working for the poor meant at most to consult with them rather than actually be supervised by them. But the Cleveland reform goes one step further. Recognizing the odds against becoming advocates for the poor when the official apparatus was on the other side, they proposed in Cleveland that the city government's urban planners actually make their plans more for the poor and the disadvantaged than for the usual commercial, financial, industrial and business interests.

Instead of making a plan in the usual manner, specifying land uses, facility locations and transportation routes, the Cleveland planners try to specify policies and objectives especially in the interests of the poor (who overlap substantially with the black citizens of Cleveland) in what they term "four functional areas: income, housing, transportation and community development."[43] It is in that context that the poor are to be given choices.

Since it is precisely those in economic poverty who are the most impotent in decisional terms, it is most important—and most difficult—for planners to contemplate opening planning so that those impoverished people may begin to gain potency. Yet the Cleveland report makes no such innovative suggestion: it accepts the traditional system of decision-making. That means that the basic choices and directions are still to be made by the politicians and technicians.

The technicians are the urban planners and others who, although refusing neutrality, do not refuse to be technicians; to provide in presumably objective, value-free ways knowledge for the decision-making specialists—the exclusive class of politicans who make the final decisions.

If this is not an innovative position with respect to decisional roles and mechanism, what about the contents of the plan? Staying with the examination of the Cleveland experience, one realizes once more that in the planning process there, the city is still read as an ensemble of parts that must function with maximum efficiency. Since the lack of efficiency is evidenced above all in relation to the needs of the poor, it is to them that one turns because public actions can satisfy them more fully by creating for them more public services.

Thus, man is once again conceived according to the usual model: a passive consumer of a given number of services corresponding to a traditional kind and number of needs. Nothing, then, appears to be really innovative in this experience and in programs similar to it.

Frances Fox Piven was very dubious that Cleveland's new planning approach would make much difference. She was dubious because, even though "a vigorous class and race oriented planning on behalf of the new urban constituency" may be possible in declining cities where impoverished minorities are becoming majorities, the decisions that really affect people are no longer being made in the central cities.[44] She refers to that as "the irony of our reformation."

What the Cleveland approach means in urban planning is the acceptance of a seemingly progressive, seemingly reasonable, but wrong assumption that will continue to make urban planning irrelevant to a substantially positive step forward. In the Cleveland planners' own words:

> Although the urban crisis to some extent affects all urban dwellers, it bears most heavily on the poor and the powerless. Crime, unemployment, substandard housing, transit-dependency and other critical problems are concentrated in low income neighborhoods.

So far, so good. But the error comes in the next two sentences:

> Solving the urban crisis very much includes solving the problems of those who live in America's slums and ghettoes. Therefore, any serious attempt to counter the forces that are making our cities unlivable inevitably entails giving priority attention to the needs of the people with the most problems and the fewest opportunities.[45]

Solving the problems of the ghettoes and slums of Cleveland, of New York, of Chicago and San Francisco, solving the problems of the most subordinated, deprived and marginalized Americans, *is* the priority. One cannot solve those problems, however, with the usual functionally divided, sectoral approach. These are problems that, though society-wide, do need to be attacked on a city-by-city basis. But the attack cannot be based on a planning approach that is either one of benevolent amelioration or the continuation of an interest-group conflict or even class struggle without attempting to open to all citizens, and especially to the poor, the decisional power-concentrating and dividing institutions. It is these, controlled as they are by small minorities, that constitute *the* basis of the urban crisis.

It seems evident that such positions as those adopted in Cleveland tend to restore the profession and the role of the urban planner in a technocratic way. These positions no longer pretend to a false neutrality as do some other positions. Instead, urban planning becomes the politicized use of supposedly but not really neutral disciplinary tools and techniques, methods and models. I shall say in a while why the pursuit of such a golden fleece is wrong and impossible.

It is not by accident that very recently most urban planners in Italy have become members of the Communist Party. They and the party profess just such technocratic views. This position also reflects the state of the profession in so-called socialist countries.

It must also be understood that, as in the Cleveland experience, in the urban planning projects of the by now numerous Italian city administrations of the left a gesture toward citizen participation is never lacking. By now, though, it is a kind of required ritual the function of which, when present at all, is none other than to enlarge the area of consensus. And, of course, I repeat, even those who take the most conservative positions no longer fail now to make their gesture toward participation. The kind of citizen participation dealt with here is even less camouflaged by pseudopolitical veils and, thus, is easier to understand.

The urbanist Marshall Kaplan's phraseology provides a significant clue to such perspectives: "Involvement of residents and elected officials in the planning process is necessary not only to help *the planner* define problems, goals, and solutions but also the help *him* set priorities."[46]

Although this book is organized around urban planning, it is important to stress its larger intention: to provide a valid tool for understanding and acting politically for the transformation of society. Therefore, it seems to me indispensable to refer at least in passing to the vast range of actions or projects that have as their direct or indirect object the increase of participation.

In Europe, even more than in the United States, so-called experiences of participation, which were numerous during the sixties, have in most instances been connected to urban-planning projects. In Italy, for example, one can even regard the so-called neighborhood councils (*consigli di quartiere*) as urban-planning projects directed toward creating further levels of local decentralized power. These initiatives were taken, first of all, by local administrations of the left and immediately after were imitated by local administrations in the hands of the Christian Democrats. Today, though, this experience is bureaucratized through a national law that specifies a neighborhood council's forms of election and the area of decisional possibility it can have.

It rapidly became apparent that the neighborhood council is not directed so much toward involving the common citizen in an authentic process of participation in regard to any matter as it is toward deepening and extending the hold of the power elite. The modern large urban areas have become more spread out and the earlier single elite at city halls was less able to control them. While strengthening the power elite, the councils also consolidate the hierarchical position of those at the level of the local government who find themselves at an intermediate or even low level of power in the overall hierarchy but always at the summit of local power. The course of the measures for opening the decisional level to parents within educational structures has been similar in its rapid process of bureaucratization and development of guided and controlled participation.

It is important to note in this regard that if some of these experiences related to urban planning have had greater life and greater results than experiences in other areas, at least in some small measure, however sporadically and temporarily, this has been in part due to the clear readiness of the planners to "surrender" or to share some of their power. A number were ready to participate in situations that didn't distinguish between planners and common citizens, between experts and those who were passively affected by the results of the plan.

It is also significant that these experiences, too, were transformed into bureaucratic facts of the kind that I have frequently mentioned. This occurred exactly when forces on the left, who had promoted such participatory experiences, became fearful they would lose the leadership, the control, the representation of the "working classes."[47]

In North America, specifically in the United States, the actions and projects that directly or indirectly touched upon participation appear in a much broader spectrum. First of all, due to historical and political reasons that I cannot dwell on here, primary interest was addressed to a specific problem: poverty, which in most cases had racial overtones, that is, it especially involved black Americans. That was of course not the case with the concentrations of large numbers of poor whites, as in parts of Appalachia.

It is well known that different theoretical models of participation, at times explicit and other times implicit, have undergirded different practical actions and projects and have been sometimes superimposed on each other. In the unfolding "war on poverty" a model was developed on the basis of the hypothesis that:

> political participation derives from social structural and cultural variables; it is not itself a means of changing the place of the poor in society or the values and action norms which the poor display. Education and other means of deliberate cultural change may be used to begin to break up the "status crystallization" of the poor.[48]

Attention was thus directed toward new educational programs, especially for children and young people. These programs were categorized under the concept "compensatory education."

Another kind of model began with the assumption that no class or no part of the population was *a priori* incapable of participating. This included actions on the local level with the collaboration of "agencies and organizations that are part of the establishment."[49] It was the task of organizers to shake the satisfaction with or acceptance of the status quo, a factor that in the model was, even more than alienation, a brake on citizen participation even on the part of the disadvantaged.

Another direction came from a very different model according to which the poor as such are subject to an entire series of deprivations, of scarcity, and of low quality of services. Such a situation cannot be overcome without giving more to the poor and, the model has it, that cannot happen through confrontations with the actual power structure.

The actions most strictly related to planning have taken root mostly from this last theoretical perspective. Advocacy planning is a good example. Whether in the context of urban renewal or model cities or in more usual slum clearance or redevelopment or conventional, routine planning processes, advocate planners, unlike Saul Alinsky and his power-through-confrontation model, have tried to avoid major contests with the powerful.

What were the results of these various forms of participation initiatives? Regardless of the particular program form or underlying model they have all had a common fate. I share the judgment of Kenneth Clark and Jeanette Hopkins:

> The campaign for massive feasible participation by the poor in the anti-poverty program must now be seen as a charade, an exhilarating intellectual game whose players never understood the nature of power and the reluctance of those who have it to share. It seems apparent that canny political leadership—national and city—never intended fundamental social reorganization. The political participation of the poor in their own affairs was not to be a serious sharing of power after all.[50]

Common citizens remained impotent, whatever the program.

We can also add another point to this criticism. When any of these initiatives took shape and it became possible to foresee a positive outcome, either someone representing government or organizations already institutionalized immediately tried to check them, to neutralize them or to eliminate them entirely. On the other hand, quite apart from the small results achieved, a critical examination in the light of what I have so far shown leads to certain conclusions that also become the foundation for a new and different kind of proposal or plan for greater citizen participation to be elaborated later in the book.

A major criticism consists once again in the fact that these programs and initiatives were based on the objective of *giving* more and better *services* or *income* to the poor. I would propose, instead, to recover the consciousness of the "totality" of the human being and of his decisional capacity. This cannot be done by giving a little more money or a few more services to those who have little or much less than others. In other words, neither our objective nor our guiding model is a society witnessing the promotion of those who are totally or very impoverished to at least the lower-middle class.[51]

In fact, modern society as a rule presents forms of alienation that do not always correlate with or act in a linear relationship to economic poverty. We find alienation, apathy and neuroses in a very large stratum of the urban population that is dominated and crushed by the models, the modalities, and the manners of contemporary life. One need only consider how vast the circle is of those who, especially in the United States, have that pervasive, constant and daily companion, the television. Despite the current style of locating all the negative features of modern life in or resulting from the behavior of the poor, we disagree.

Thus, one must carry out any innovative program on at least

two fronts: the first among the economically dispossessed and disadvantaged and the second among those who, while not in the same category, are alienated and deformed even if they do not recognize what has been taken from them as human beings. In any such action, the redistribution of power and economic resources, until now artificially seen as two distinctive, separate actions, regains its natural unity.

Finally, as I have suggested in Chapter One and will continue to show in greater detail and depth, I envisage urban self-management as the proper vehicle for extensive citizen participation in the future. Urban self-management does not mean that only the poor would manage activities pertaining to them; urban self-management means applying the principle of self-management to all institutions, thereby opening them up and breaking their present hierarchical structure.

4

NOTES FOR AN ALTERNATIVE CITY AND SOCIETY

The Meaning and Role of Institutions

I sketched earlier the characteristics of modern urban society, its problems and its current and not pleasant general tendencies. I stressed the point that if we do not bring about radical changes in our understanding of society and in the means we use to deal with it, those general tendencies will become overpowering forces.

In the third chapter I showed that urban planning is one of the means usually used to try to improve sociophysical organization, if not to modify it and to heal its ills. Actually, though, urban planning is likely to have opposite effects. I hope that it has become evident, and to nonspecialists too, that urban planning is based on those same basic assumptions that make modern society increasingly less workable and less livable.

For this very reason, I asserted that urban planning and its instruments must be changed in a far-reaching way. Marginal variations, retouching, adjustments, updating and the preparation of more specialized methodologies are not enough. Nor can we look to new, sophisticated professional tools. Consequently, all of the so-called new proposals that involve no more than such things are neither sufficient nor even positive.

Their common matrix, and the source not only of their inadequacy but also of their negative character, comes from the fact that they are the result of reflections "on urban planning." Instead, we must reflect on society, on urban society, on transforming urban society and from there return to the processes of urban planning.

Let us, then, assume this starting point but not with the idea of producing a completed theory of socio-urban development or a practical manual for the transformation of urban planning. Intellectual energies should be spent in seeking to elaborate new concepts for understanding current society in its totality. That is, we must find concepts to substitute for traditional disciplinary and even interdisciplinary or multidisciplinary perspectives. Perhaps at that stage we will have arrived at concepts only partially shaped and minimally useful, but they, in turn, will be able to serve as the point of departure for other people's contributions.

It is my conviction that the human being continues along a route of ever greater liberation, of the ever greater realization of his humanity. This, however, does not seem to be true of the moment we are now passing through. Yet no one in any other epoch has been given his freedom and humanity *gratis*, nor can we expect it to happen to us. The realization of freedom and humanity is the result of a struggle that cannot but have reverberations throughout the entire so-called social system. Those responses may provoke successive improvements or result in worse and less human conditions. But it is in the nature of human beings to try to recuperate and improve. In order to give a concrete example of my position in that regard, I will discuss industrial development.

I do not mean to lament the happy age of healthy poverty in the agricultural-pastoral-rural world. Yet, the advantages stemming from industrialization have also brought problems. And it cannot be said that the type of industrial development that we have known in the past, know now, and that is proposed for the future is the type that can continue to be beneficial to man. It is proposed for the future because of its understandable and known characteristics.

Thus, to be opposed to the present forms of economic-industrial development does not mean to stop development altogether, to desire poverty and misery forever for those people now in such conditions. Rather, it means to opt for and to seek that type of development for which humanity need not pay a great price in terms of dehumanization. Given the class reality of capitalist society, the new "underclasses," more inclusive than the traditionally subordinated classes, pay disproportionately for the present forms of this development.[1] Perhaps in this case the person who says "no" to the reproduction of the present state of affairs is more positively oriented than he who does not even attempt a critical examination of it and simply projects and supports it for the future. In fact, faith in industrial development as we have known it is often simply a distrust of man's ability to construct a better alternative.

With respect to the possibility of constructing a different society, based on different parameters from those that cause the aberrations, distortions and distressing problems of present society, it seems necessary to make one observation. Until now, my criticism of the fact that there has been little or no research into alternatives has, above all, been addressed to urban planners and urban planning. My criticism, however, is not intended only for them; the disciplines concerned with society and their respective professionals share, at the very least and more or less equally with the planners, that responsibility.

And if we want to maintain the usual hierarchical organization, one that technocrats prudently accept, we must place the experts in governing, the politicians, at the top of all these variously labeled specialists in the organization of society. The politicians, in other words, bear a major share of that responsibility.[2] Their lack of imagination becomes most obvious at a moment like this when it can be said that no country is exempt from profound social problems, although one can exclude those frozen into dictatorships that put them outside the processes of social change and evolution.

The presence of such problems and their intensification should have provoked a search for new paradigms with which to confront them. Instead, hardly any of the present notions, methods and instruments go beyond traditional economic measures or beyond the creation of further protective structures (modern police forces and the like) or beyond an enlargement of the field of action of those professions that serve as safety valves for social ills.[3] There is almost no glimpse of a new society in the United States with Carter, in Italy with a rising Communist Party influence, or elsewhere in the more advanced urban-industrial societies of the world.

To enter into the substance of our argument, however, one way to proceed is to ask what exactly do we mean when we speak of transforming society, of a new society? Better yet, we might ask about those very basic terms that I have been using so far: human being, institutions, society and the city. In this case, too, my purpose is not to give new definitions for terms; rather, I would like to explore certain aspects of them that will permit us to define the objectives that we wish to pose to ourselves when later in this chapter we speak about alternative forms for the organization of the city and human life in the urban environment.

Let us begin our discussion with a short story. There was a man who wanted to shorten the time needed in going from home to work. To do so, he first built a bicycle. Then to further decrease the time he bought an automobile. But the travel time can be shortened even more, he thought. So he built a vehicle so powerful that one day he lost control of it and ended up smashed against a wall.

At first sight, some will undoubtedly consider this story to be an image of the very history of humanity that, for example, risks destruction by playing with nuclear power: it is born in "legitimate" discovery, then nuclear armaments are devised in the desire to defeat fascism, and subsequently increased risks arise from the desire to exploit the nonmilitary advantages of nuclear energy.

Actually, however, in the case of our protagonist, the commuter who wanted to build a more powerful machine, we must seek the cause of his final disaster in his very self, in the objectives that he set up and in his erroneous evaluations. In the case of "humanity," however, we know very well that when using that term we are not speaking about a single acting subject. On the contrary, by reading the most ancient as well as the most recent history, we see that often that person who within the category of humanity takes certain decisions and accomplishes certain actions is not the one who pays the consequences of them or, at least, feels them the most. But there is more to the imprecise connotations of the word "humanity."

Let us return to what we spoke about in the second chapter. There we saw that the city and urban society seem to have become less and less suitable environments for man and for the expansion of his human qualities. Institutions seem increasingly ineffective there. In many cases they have become so gigantic and so intricately interconnected that man, the common human being, actually seems a pygmy in comparison, a miserable and inadequate being when seen in relation to them.

Even in this respect there are those who react by speaking of the "necessity" of "risk factors" that "humanity" has "had" to assume, "tributes" to be paid for the great conquests "humanity" has been able to achieve only through such urbanization and institution-building.

Again, the term "humanity" serves to cover, to mystify, the reality of some men having taken advantage of a certain type of development, of certain organizational forms of society, at the expense of others. It is my intention to consider this in some detail by examining precisely the matter of institutions.

Many people, the vast majority, assert or believe or assume that institutions were produced by men in order to facilitate and regulate their relations with others or to facilitate the satisfaction of their needs. It is through institutions that the great values and services such as justice, education, public health, and so on are guaranteed.

No matter what particular definition of institutions we take, we see that they are treated as superpersonal and suprapersonal entities having the property of existing in and through time, beyond the limits of each single human life or even of generations of its own members. These institutional macrostructures are considered to be entities devised in order to satisfy a part or a series, but in no case the totality, of human needs (here again is the well-known model of the social human being who possesses a series of needs).

Those who study institutions usually divide them into subcategories: economic, social (dealing with the family, etc.), political (dealing with the government, etc.), cultural, recreational, religious, and others. Society itself is regarded by some as the largest of the institutions, the one that comprehends all the others; other thinkers consider the government to be the most important and the "highest" institution while in classical Marxist theory economic institutions are considered the basic ones. Society may also be regarded as an entity whose elements are the other institutions. In more recent times, it is also possible to read in this definition the sense of society no longer seen as a sum but as a "system" of institutions.

The development of "systems" thought has not only focused interest on relations between institutions; it has also introduced the idea that institutions are not fixed entities invariable in time but are, instead, continually subject to modifications—just as is society.

Beer concentrates his efforts on demonstrating that institutions are not fixed entities; institutions are complex, thoroughly dynamic systems.[4] His basic argument is fascinating and persuasive: institutions are dynamic because they are characterized by the presence of people. In other words, men are the basic elements, the true essence, of institutions. Institutions, indeed, as they are commonly conceived, as enduring superpersonal entities, actually do not exist. Although Beer does not draw just that conclusion, we do.

But is this true?

Without a doubt many people, including other systems analysts, would deny it. For many people institutions are an ensemble of fixed and abstract aspects (laws, regulations, rules, customs) and at times physical aspects (facilities, buildings) that of course are made to function, are operated, by men. Institutions are important, however, and take their essential character precisely from their nonhuman aspects, which aspects constitute the continuity, the enduring fabric, of society. So goes the usual argument.

Indeed, even though institutions, too, are subject to changes, the rhythm of their variations is not comparable to that of a human being. They are relatively time-invariant, that is, in relation to people they can be considered invariable, except for the transformations deriving from exceptional events. Men pass away, institutions remain. Institutions may change; people must die. This is the basic assertion.

Such an assertion, however, is not without some contradictions.[5] Its advocates are forced to make some amendments. They must propose that even though institutions endure, they can take on different forms or have different properties or interpretations in different social situations and in various historical epochs. Thus, they can explain that in a nation that has passed from a monarchy to a republic, for example, institutions from the preceding period can be made to function substantially unchanged. That is, once again we have superhuman structural variables that make it possible for superhuman structures, that is, institutions, to function with continuity even though they may be operated differently.

The interpretative hypothesis that I propose instead is to set aside the classical image that institutions exist when rules and laws and physical apparatus exist and to substitute for it the notion that an institution exists when, and only when, there are human beings to make it function. It is human beings who interpret and reinterpret those rules and laws and human beings who use those physical facilities. It is mystical, in fact, a fiction to regard laws, rules and the like, and physical structures too, as having any meaning aside from the meanings assigned by living people even if these meanings assigned by living people are appropriated from the meanings once assigned by people now long dead.

If we consider what happens in the nonmystical reality of everyday life, we realize that the things that occur belong basically to one of these types: either people find the effects of an institution to their liking and therefore support it basically as it is or they dislike its effects and seek to change it.

When people find that it is in their interest that they continue to enjoy certain effects, results that they have had before, through an institution, they seek to prolong those effects. If external circumstances vary, people may seek to vary the interpretation and the use of the laws or norms that regulate or characterize the institution if that is the way to obtain the same desired result.

If this is not possible through variations of interpretation, they may seek to modify those laws and regulations. In either case the possibility of success does not per se depend on the laws and regulations but on the fact that other people either dispute the interpretation or modification or, instead, accept it. Actual everyday and historical experience teaches us what and how many different interpretations can be given to institutional rules, regulations, laws, doctrines. That is the case even if on balance the evidence is that general patterns continue over long periods of time in regard to institutional operations and, therefore, institutional character.

When people find that it is in their interest to modify certain effects, results that they have had through one or more institutions, they have the possibility of trying to change the use of existing institutions and their attributes in a different way or of trying to modify them. Outcomes depend on what effects and results other people or other groups are able to accept or suffer, or not.

The view that the essence, the nature, of institutions is man rather than something abstract or superior to him, surely makes a small Copernican revolution necessary. Some people will resist it, others will oppose it, and still others in their rejection will counterattack. Few institutional professionals can accept this view.

In this moment in which it appears clear, beyond dispute, that institutions are not functioning very well, that they are oppressive and rigid, we see that attitudes toward them vary. There are those who, in rejecting the conclusions derived from the man-as-essence analysis or the analysis itself, hold that the crisis of institutions comes not from inflexibility but from the opposite: that they have become too permissive, too flexible. The proposed solution is that for the "common good," for the general public interest and order, institutions should be reinforced, strengthened and made less flexible, particularly those that seem to specialize in producing social control such as the police and other institutions of criminal justice, the family and the schools.

I would like to dwell a little longer on what it means to grant to institutions a superhuman functional status, a reality, an existential reality, apart from their constituent human beings, and to see how this concept functions today in the life of modern man.

First of all, one no longer obtains his rights and services only from the fact of citizenship in a country. It is accepted that he has various rights only if he is a member of one specific institution or is entitled to various services only if he belongs to another. In short, human equality is not only fragmented into class categories but also broken up into a myriad of institutional categories and affiliations evidenced by membership cards, identity cards, and so on. Even more: human consciousness itself is "institutionalized."

Let us give an example. A person on the board of directors of a given private company or public agency takes a certain, perfectly legal, action, which, let us suppose, causes damage to others, financially or otherwise. Those who are damaged may find this just, moral and even proper once such action is justified by the fact that it has been taken in the name of the corporation or agency. In fact, it is often regarded not as the action of persons, even as agents of a corporate "body," but as the action of that body itself, treated in much of Western jurisprudence as a person.

The fact is that these days fewer people than ever are still concerned about, or see, or think of the actions of institutionalized groups as being actions of particular people who act in their own interests of particular other people.

There are people who criticize specific institutions or specific aspects of institutions but defend the principle of institutions. They might respond to my example by allowing that there are some institutions or "subinstitutions" that actually constitute instruments for masking the exploitation of man by man and that these must, therefore, be changed or even eliminated. But there are other institutions that are useful because they truly correspond to the satisfaction of human needs. These must be maintained even if also somewhat modified.

Further consideration convinces us that such assertions, however plausible at first glance, are wrong. All institutions are organized on the principles of permanent hierarchical organization. Such organization must sooner or later be repressive and alienating. The very acceptance of the concept "institution" involves the acceptance of its mythical superhuman, suprapersonal significance. This inevitably leads to the logic of measuring results and performance on general abstract principles rather than confronting the necessarily specific effects of people on other people.

Furthermore, the fundamental aspect of institutions that justifies their specialized partialization and separation as being the satisfaction of human needs also characterized as specialized and separate is negated by the exclusive nature of institutions. In fact, as soon as an institution is created, two categories are established: those who are qualified to be in it and those who are not. In other words, a differentiation is created, an inequality between those who can enjoy membership rights and uses and those who cannot. Needs must immediately be satisfied on a differential basis to the extent institutional membership is meaningful. These boundaries of, these limits to, institutions also cause de-formation in both consciousness and practices. Educational institutions may be used as an example. One way of limiting or bounding them is to limit the educational process to the first period of man's life. Yet, it is obvious that the educational process extends over one's entire lifetime.

Traditionally, institutionally, education has also been bounded by or limited for the most part to schools. But the factory, the farm and the family are as much places and occasions of education as is the school. This, though, would violate our institutional thinking, our institutional consciousness, if admitted, not as a platitude as it is now, but as a basis for action. Such admissions of simple truths would wreak havoc with the kinds of "market-sharing" agreements that minimize conflicts among institutional hierarchies.

I hold, therefore, that the distinction between innately or intrinsically good and bad institutions does not exist; it is wrong to think that some need to be destroyed while others need merely to be improved. I do not agree, however, with those who want to destroy all institutions in the name of either anarchy or total individualism.

I must insist that the human being is such only in his social nature, that is, insofar as one is a human being among others. This, of course, immediately requires forms of organization that also certainly assume a binding and sometimes a commanding character. I am well aware that with respect to human organization one must also face such problems as laziness or people who use frequent headaches as their excuse for nonperformance or whose moods are highly volatile and make life difficult for others. But it seems plausible to think that people have enough "natural" wisdom to cope with such problems. And of course many problems, such as frustrations, aggressiveness, vices and defects, rather than being the result of original sin, are actually the result of a distorting and false morality, and of the pseudovalues of the social system into which people are indoctrinated—by other people.

All persons are not of the same character. Thus, modes of so-called social control will be needed for an indefinite future. But that means only that some people will be guided or controlled by other people and not by superhuman and abstract entities called institutions, whether those be government, society, the schools, or whatever.

To repeat, it is a mystification of institutions to conceive of their essence as superhuman because, actually, it is always human beings who act. Through this disguise, however, ways of operating are made possible that can and do become inhuman and antihuman.

I would like to cite an example from the field of electronic computers. Those who, as laymen, have the opportunity to work with or near computer experts first marvel, and then perhaps become accustomed, to hearing phrases like the computer "is acting up," it "does not want" something or it "has decided" something. The phrases seem innocuous, but they are not. The expert progressively assumes the halo of the high priest, the one in the position to communicate with the "god" computer, the only one capable of pacifying it, or of obtaining or influencing *its* decisions.

This may reach the point where one is nearly convinced that it is true. At times the expert himself acts as if he really is convinced that the "god" computer decides while he, the expert, just influences, despite the truth, known by all computer experts, that decisions are only and always made by human beings. We have greater difficulty than ever before in escaping the trap such experts prepare for us when, faced with an error in some office or with a totally human, erroneous decision, we hear the resolute and nondebatable reply, "It was the computer" or "The computer isn't working properly."

This new and growing mystification produces two negative effects. On the one hand, it raises a new, science-fiction kind of barrier behind which human actions and manipulations, right or wrong, can entrench themselves. On the other hand, it prevents rather than allows understanding and use of computers for man's good in a nonmystified, human and significant way.

The mystique of institutions functions in much the same manner. One may conclude, as Beer does, that institutions must be redesigned. Should we aim to redesign or to destroy them by substituting other organizational forms for them?

This is merely a trivial semantic problem so long as in redesigning or destroying by substituting other forms they become, in contrast to present institutions, not "exclusive" and not superhuman. That is, the consistency and rationale of their workings must come from human values and they must be inclusive, open to all.

Such institutional opening and demystification will not be easy. There are concepts, Marcuse says, that have also become biological characteristics by being inculcated into the individual through widely different forms and from generation to generation.[6] Once such concepts are no longer recognized as valid, they do not, therefore, immediately disappear. Generations are probably necessary before one can be totally free of them.

To possess the tendency to act as if institutions were superhuman is to accept associated hierarchical stratifications; exclusions; impositions deriving from abstract, unreal principles; and fragmentations. These can even survive an eventual categoric decision to abolish institutions. For such reasons I am not proposing that. I do not think this should be the starting point.

The abolition or withering away of institutions and the defining of new organizational forms constitute, instead, a process that can occur only by being identified with the process of man's de-alienation. This, we know, does not only amount to a problem of economic de-alienation.

Alienation, especially today, cannot be characterized simply by economic exploitation, not even in underdeveloped countries or in those areas where there are still such brutal facts as poverty and hunger. The masses of the Third World suffer a more encompassing alienation from all kinds of exploitation by elites in and outside their own countries.

In other words, there are different and more refined lines of demarcation between the dominant and the dominated, between the exploiters and the exploited, than the economic indicators we usually use. These are the considerations that lead us from the discussion of institutions to speak again about the city.

Indeed, the city is not an institution in itself but a spatial location that can be bounded; a definable place, where institutions are located in geographical space.[7]

Thus, although spatial location is not a part of the definition of an institution, the city is a kind of "container" of those portions of an institution that exist within its boundaries. Moreover the city is the place in which human relations in the modern epoch have increasingly assumed the mystifying sense of being relations between institutions.

As the French sociologist Castells writes about modern urban development: "On the one hand, the segmentation and utilitarianism of urban relations cause the functional specialization of activities, the division of labor and the market economy; on the other, since direct communication is no longer possible, the interests of individuals are defended only through representation."[8]

It is in the city that maximum visibility is given to the fact that man does not receive according to his needs nor even according to how hard he works, but according to the institutions to which he belongs and to the myriad of status and influence relations involved therein. In the acceptance of this ever more extensive and pervasive specialization, he who has more institutional positions from and in which to operate not only often has more power but also has a certain moral condition. By this we mean that it is through institutional performances and identities that a person becomes conscious of his rights, that he has legitimate claims on others. Just so do others come to accord him a moral status.

I do not want to deny that either always or even often the rights of dominated classes and groups are collectively assigned and shaped by such memberships and identities. But the meaning of these rights, their reality, takes shape in and through how people perform in the institutional matrix. Nor do I intend to argue that people ought to make claims merely in individualistically personal rather than in collective actions. I believe, though, that it is necessary to be more specific here in order not to be overwhelmed by the suprahuman mystical archetypes with which our minds are encrusted whether those be of institutions or classes. "Collective" is one such mystifying adjective when it moves from a meaning of people together toward the connotation of superhuman collectivity.

In the present situation, for example, it is opportune, even necessary, for a worker without a place to live to seek the solidarity of others who are in a similar condition in order to obtain his rights (or "privileges") as a human being. This situation is, however, a pathological one. The desired situation, for us, is not that in which the right of a man to have adequate housing is recognized because he "belongs to" a category, a class, of a large number of workers. Rather it is one in which his need and rights are recognized because he is a man, not because he is forced and forces himself to accept a partialized, nonhuman, institutionally defined categorical functional identity. The importance of keeping this constantly in mind seems to me to be fundamental.

It permits us to proceed immediately to another consideration.

In order to transform the situation of the city, one must find a way different from the indirect one of improving institutions, which presumes that once institutions are improved, they will then improve the life of the city and of men. Such a presumption is an erroneous foundation stone of modern thinking.[9]

One must, instead, pose the problem of how men's lives within urban space may be improved directly. In other words, this means asking first of all what changes can be considered unquestionably as improvements of human life. Evidently, it is not a simple problem. The question, then, of whether such improvements ought to be made inside or outside of the contemporary institutional matrix is the second-order question.

Recently, for example, an American newspaper reported that NASA proposed to other agencies of the United States government to resolve the food shortage problem for poor, elderly people by producing a large quantity of highly nutritive synthetic foods of the kind used by astronauts. Besides being inexpensive, the foods are highly imperishable so that minipackages could be sent cheaply by mail, simplifying the so-called delivery-of-services problem.[10]

Some will surely consider this a marvelous technological success and a marvelous new service. Others will question whether this would be an improvement of human life and I number myself among the skeptics. And I do not think such a new service from the institutions of the post office and those of social welfare will even indirectly improve the cities or contribute to the transformation of urban society. Transformation means something else; it is not easy to define but it is profoundly different.

Perhaps the most useful way to define alternative lines of action for society and the city is precisely that of asking ourselves for what kind of man, what kinds of persons, we want this city and this society. This means, obviously: what type of man can be simultaneously the object and the agent of these transformations?

The attempt to clarify this might constitute the most important tool with which we can provide ourselves for formulating adequate ideas and proposals concerning urban organization, as I proposed doing at the beginning of this chapter.

Toward What Kind of Human Being?

In the preceding chapter I examined the traditional processes and results of urban planning. Reference was often made to the objectives that urban planning ideally or theoretically should have and the model it should use to achieve them. The model was characterized by two requisites: the social quality of urban space; and equal access to the use of such urban quality.

Earlier, I used this model purposely without defining it in a precise way, content with what the terms "quality" and "equity" or "equality" usually mean for each of us. Now, however, it is important to examine the substance of the terms, especially so for our objective of defining the type of man for whom urban planning should operate.

Let us begin by exploring what we mean by "the social quality of urban space." Certainly, the problem of defining it and uncovering all of its aspects in a way that will satisfy everyone is not a simple one. But what is not a satisfactory approach is the economic approach. One cannot define the city and society as factories whose quality of functioning is measured by the quantity and quality of the production or outputs. (Of course, even in the case of factories, using those judgmental criteria while ignoring such dimensions as working conditions, personal growth in the job and participation in decision-making leads to problems not unlike those which we find ourselves facing today regarding narrow functional conceptions of city quality.)

As this dismissal of the economic approach shows, when it is difficult to establish the parameters of what it is precisely that "we want," which often occurs with us human beings, it helps to try to think about what "we do not want." So it is with respect to those aspects that concern man as a psychobiological-socioeconomic entity in the city. We know that besides satisfying needs that are called essential or basic, we also do not want man to have to face such urban problems as air and noise pollution, chaos and unsafe streets.

We have seen that the negative conditions to which man as a social entity is subject in modern society are most acutely manifested in the urban environment. Those negative conditions originate from and are deeply influenced by the division and specialization of labor and the divisions of life. We have also seen that the organization of the city according to institutional demands and domains is a central factor integral to these two phenomena and one that shapes their expression. To this are also related the urban ills present in the forms of privatization, neuroses, cynicism, segregation, apathy, and mental illnesses as well as crime and delinquency.

Other aspects of urban life that bring uneasiness or malaise include the emphasis on money, consumerism, the fetishism of fads and products, and so forth. The major difference between these types of manifestations is that some, such as violence and delinquency, blatantly break with the rules of a society so organized and provoke a reaction. Others are merely the logical although extreme results of the workings of the specialized institutions of divided labor and life. Mental illnesses also tend to be regarded with fear for their ability to be manifested in "antisocial" behavior. The other so-called urban ills or malaises are simply endured by the great majority of people, who feel vaguely unsatisfied while usually accepting as inevitable aspects of the "model of life" of this society and/or as personal failure.

The fact that manifestations of such conditions as delinquency and violent crimes can be measured at least roughly (the number of cases, type of crime, etc.) might lead one to think that a simple measure of the quality of urban life can also be articulated. In this view, life in the city is improved, is better, when there are fewer thefts, fewer cases of violence, fewer killings; in other words, when urban life is more ordered, more regulated and more tranquil.

History has taught us that it is not so simple. Indeed, there are numerous examples of situations in regimes of various historical epochs—in the recent one, the Nazi regime—in which order and discipline were not the fruit of the self-governing and healthy social life. On the contrary, they were the result of a system of control that protected itself through a freezing of social life, which gave an external appearance of order.

Another example is the American middle- or upper-class suburb. On its ordered and protected chessboard, crimes occur much less often and less easily than in the chaotic urban center. Many of those squares, however, contain buildings in which persons live neurotic, psychotic or depressed lives. They certainly do not express the joy that one might be led to think of as the positive, reverse side of the minimally negative conditions in suburbia. And of course such persons are not deprived, comparatively and often absolutely, in terms of material well-being.

What we have just said suggests how we believe that it is possible to define positively the social quality of urban life—not as the minimization of negative conditions, whether pathological or accepted as normal, but as the maximization of at least the following conditions of the human being; sociability, creativity, personal vitality, interest and pleasure in participation, a variety of respectful social relations, a multiplicity of interests.[11] Several of these require care and concern for others and all are rooted in an image of people as human beings living in a respectful, secure freedom.

These can constitute for us the first traits, a preliminary model, of social man that in order to distinguish him from alienated and divided modern man, we can call "total" man. Our more inclusive model says that the social quality of urban life varies as that quality is congenial or adverse to the growth and development of total people so defined.

If we examine these conditions one by one and all together, we will realize that few or none of them can be influenced by the improved functioning of institutions nor can the conditions thrive when subject to the usual divisions and partialized characterizations to which modern man is subject.

This model's concept of creativity does not refer to the creative expressions that modern man practices as a "hobby" during his free time in order to compensate for his daily routine of work. Nor does it embrace much of the life of the designer who must commercialize his creativity by teurning it into professional work, seemingly free and creative but in reality bound by the most rigorous impositions of consumer society. Much more than the usual institutionally restricted sense of artistic quality is intended here. I am referring to the gamut of expressions of versatility and ingenuity that are denied to so many people so often—and unnecessarily denied. Those who must do office or factory work ordinarily do so in a machinelike manner that conforms to the uniform standards required by norms of efficiency, profitability and hierarchical accountability, constraints imposed and maintained by men and justified and mystified in terms of necessary institutional requirements and such norms as that of obtaining or preserving excellence.

So, too, when speaking of sociability, I am not referring to the stereotyped smile of the salesman who considers it a professional instrument. If it is necessary to the smooth functioning of his business rather than a spontaneous result of experience with other human beings who are his clients, then I would not consider it sociability.

The conditions of which I speak seem very natural and normal. Someone may even ask if they are in reality significant enough to be posed as goals. Others may question whether institutions may not be reformed so that they can improve such matters.

To answer the first question I would refer again to the example of the astronauts' food tablets to be sent through the mail to the elderly. It is merely a grotesque example of how we forget the simple human needs of people in the modern urban society. The elderly need more than supplies of proteins and vitamins furnished as if they were cars to be filled with gas. Daily life is full of such examples of inhuman solutions to human problems, and many of our superhuman institutions provide those examples.

Let us look more carefully, though, into the second question, why we cannot obtain a progressive maximization of the characteristics of high-quality social life by improving institutions in the cities.

We have noted that institutions actually constitute instruments that some men use to condition, control and dominate, or at least manage, others. Precisely for such reasons, institutions take man apart and divide one man from another by categorizing men according to their various parts. The chopping into pieces is justified on the basis, first, that men are in different domains according to institutional distinctions and, then, that they perform diverse operations and have different consequences or functioning within institutions.

By upholding institutional mystifications, for example, public services that may take over from private corporate exploitation retain aspects of alienation for the worker who produces the services and who is alienated from his product just as is the citizen who receives the service. But the aspects that make a service alienating for those who receive it are somewhat different from those pertaining to the service providers. First of all, the giving of a service is not an exchange, a reciprocal, mutual relation but a unilateral relation. Furthermore, it does not take place on a basis of parity. He who receives it has a need and, therefore, is in a position of inferiority with respect to the "specialist," however low level, who can satisfy it. This need becomes a condition of inferiority and of dehumanization precisely because it is extracted from the totality of the human person. The one "served" is not regarded in his totality as a human being, but in that partial aspect that requires the service, as a part to be serviced.

If an institution could be modified to the point of really changing such alienating relations and conditions, to involve the users of an institution—the consumers of its services—in participating in its operations, this would become per se an enlargement of its limits as an institution, putting it at the level of human beings and no longer above them.

Similarly, if in an institutional setting people began to treat each other as total people through a broadening of the institutional spectrum of action it would no longer be a specialized institution in its usual modern connotation.

Let us take educational institutions for an example. We might think that they are in a position to improve many of the relevant conditions of social life in urban space if, for instance, they would engage in processes for the continuous self-education by persons of every age. That would have to involve the appropriate objectives, methods, tools, places and forms being created and reevaluated by those same citizens. But the specialists, the educators, would have lost their special social status in the process.

Such educational institutions would no longer be the narrow and limited institutions of today. They would not even remotely resemble present-day schools. I will return to the whole subject of education and the development of knowledge later.

The qualities defined earlier as positive, as good, were conditions to be maximized not by single human beings but in an associative life with others. These qualities do not find much scope in today's society but not because men are bad. It is because the social fabric, the person and, consequently, the city are structured, are shaped, on the basis of the criteria that undergird and permeate the distorting divisions of labor and life. These criteria operate neither in the interest of the single human being, nor of people taken collectively, that is, society. In a general sense, the criteria even act against people despite a seemingly popular acceptance. They generate and increase social injustices despite the construction of welfare states or even socialist societies.

At this point one is forced to ask how, in reality, can the already very great divisions of life and work not only maintain themselves but even become deeper in this phase of industrial society? How can such divisions threaten to become even deeper in the next phase? It is an important and fundamental question, to which no simple or brief replies except simplistic ones are possible. Let us try to see the situation by starting precisely from the concept of injustice.

What are the major social injustices connected to the division of labor and life and why are they maintained? I think it is quite evident as to who benefits and, therefore, the question is in one sense superfluous. Marxist analysis teaches that the capitalistic system functions by dispossessing the working class of its products. We have also seen that an accurate reading of Marx can lead us to conclude that "products" do not refer only to economic ones. In other words, economic privations and deprivations are not the only privations and deprivations that characterize modern society. The rich and well-to-do can be, and are often, poor!

A large set of injustices comes from the continuous expropriation of decisional possibilities to which the mass of citizens is subjected. The distinction between those who govern and those who are
governed, between the one who plans and the one who is planned,
constitutes the basis of contemporary society. Such a distinction is
only weakly correlated with wealth, income or even class in orthodox
terms. By that I mean that while it may be necessary to possess top
economic position or power to avoid decisional alienation, that
power alone is rarely sufficient. Such decisional potency-impotency
stratification has become even more significant than the division
between he who owns the means of production and he who only
owns his own labor-power.

In fact, it is well known that "power" in contemporary society
is not necessarily always in the hands of the owners. The most profound line of separation is, instead, precisely between the one who
makes decisions (whether he be the owner, the manager or the tech-
nocrat-expert) and the one who is obliged to obey.

To be sure, this has always been the line of demarcation. Indeed,
in no historical epoch was the ruling class narrowly constituted by
those of greatest wealth. There is no doubt, though, that in the
present epoch it is no longer a matter of a few able politicians being
in a position to influence those in key positions in the socioeconomic
system. Given the present organization of society, access to and pos-
session of a certain kind of preparation and a certain type of information and skill are decisive factors in determining whether or not
one can belong to or even be the rewarded agent of the privileged
class of "decision-makers."

Since this expertise and information are not open to all, this
constitutes the basis of the kind of discrimination and social injustice
of which we are now speaking.

Let us, in fact, briefly look at what I have vaguely called "information" or "knowledge." It is an elemental reaction for all of us to
think it is connected to the educational system. We usually assume,
for example, that there is an equation between the years a person
dedicates to "learning" within educational structures and the accu-
mulation of knowledge and the capacity to manage, at least if the
person is a good student.

It does not happen this way. On the one hand, the school, and
especially the public school system, is constructed in order to channel
varieties of students and homogenize them according to preestablished
parameters. On the other, schools are outside the circuit of real-

world knowledge and outside the real world. Thus, real-world knowledge of the character of decision-making is learned in another part of the system, not in that of public education.

In fact the evidence is that top decision-making positions and even those depending on skills acquired in schools depend more on knowledge acquired in the home and in the extraschool social environments, on the one hand, and in the work setting, on the other. The fact that people in top management, for example, are not essentially formed within public educational structures is another of the conditions through which the dominant decision-making class seeks to guarantee its stable control.[12] It is, in fact, only the elites who learn or come to understand, consciously or semiconsciously, that people and not institutions make decisions. That rarely is learned in a public school system that is itself part of the modern false institutional consciousness.

I do not intend to imply, however, that in order to overcome this injustice one must somehow give equal opportunity of access to the processes of the formation of the elites to the person of working class background so that he, too, may be given the chance to become a managerial decision-maker. Such is the traditional objective of a good reformist perspective. Instead, I mean to suggest that injustice exists as soon as there are classes of those who make decisions and of those who must undergo or submit to them.

The fact that decisions, once taken, need to be carried out does not mean that the world must be seen as necessarily divided into those who follow and those who decide, specialists in their respective roles. Indeed, such a division is absurd if one considers that even a specific decision in the most detailed program involves continuous changes, that is, new decisions concerning its execution. Those who have participated in the initial decisions should be "more capable" of carrying out the new ones. They are in a better position to modify them appropriately when that becomes necessary. Thus, we should either have a small number of people monopolizing all phases of decision-making or open all phases to more people.

Of course, someone will tell me that it is difficult to imagine the social justice we are talking about, the chance for *everyone* to participate in *all* decisions. It could be said that many people—a very large proportion—are not in a condition to take part in them. I can and do accept the fact that presently few participate in various decisions, except to obey, but not that only a few are capable of such participation. This objection evades the real problem: whether this should be a goal for achieving justice. My answer is affirmative. I have dealt in Chapter One with some of the problems to be faced in opening the decisional processes and the tools that may help in this regard and I will later return to these matters.

I made a reference to information as a necessary, if not sufficient, element for participating in the decisional process. But today information is also something to which people have access in non-egalitarian forms and, it therefore, becomes a factor of discrimination. I am not only speaking about the manifold "impossibilities" of access to certain kinds of information "for reasons of state" or "for national security" or even of prohibited access to "business secrets." I am referring to the entire manipulative process of information control and management that is another crucial element for preserving the present hierarchical and exclusive decision-making system in all institutions.

It is evident that these types of injustice are closely related. One cannot rely for a remedy on the liberal slogan "knowledge is power" because it is impossible to imagine how knowledge can be distributed equally by elites that have interests in its concentration, in its unequal distribution. Nor can the present situation be transformed without simultaneously attacking the several basic aspects by which social injustice is expressed and articulated.

In other words, we cannot pose as our goal for achieving justice the image of a society where a ruling class, active, honest and fair to all the social elements, makes institutions function in an efficient, ordered and egalitarian manner. That is a mystifying, mythological, impossible image.

The image is attractive because it conjures up, however vaguely, conditions that seem better, much better, than those existing today in capitalist countries. I do not, however, mean to imply that I see a more advanced solution in contemporary socialist countries. Only some of them can actually be considered to be in a state of transition *toward* socialism. Others are permanently bound to be not basically different from contemporary capitalist countries.

It is not accidental that the socialist countries that are possibly in transition toward truly different forms of society are precisely those in which institutions are seeking to broaden their decisional processes. There are countries, in other words, where experiences of self-management are at least attempted. We have seen some of the positive results of these experiences in Yugoslavia in the first chapter.

With these initial observations it seems to me that we have made a small step forward in defining "quality" and "equality" as objectives toward which we must move if we want to transform society and the city. And if we agree on these initial concepts, we have an instrument in common that permits us not only to elaborate proposals but also to see and evaluate which present approaches really move in the direction of these transformations.

In order to be clearer, it is perhaps necessary to pause for a moment over the terms that, according to the context, may be used to express such judgments. First of all, reality is never static. The situation that exists in a given moment constitutes a change with respect to a preceding situation and in turn constitutes the premise or input for the future.

I use the words "premise" and "input" because I want to underline the fact that in no moment is this process predetermined. But I want to do more than draw attention to the fact that human processes, in fact, are recognized as probabilistic ones by all the traditional disciplines that study them. I also want to stress once more the fact that human actions can produce in every moment substantial changes in a process that is by its very nature already one of change. That is true even when the changes are re-creative, repetitive, conserving.

These changes need to be defined. With reference to our previously identified requisites of the social quality of urban life, the changes can be reduced to these three general ideal types:

1. transformations and modifications in which the actual or projected changes are such as to worsen the situation of the socio-urban system insofar as they worsen the conditions of social quality and equality—in this case the changes are *involutions*;

2. transformations and modifications in which the actual or projected changes are such as to increase the functionality and efficiency of some institutions but not such as to effect improvements in quality and equality—in this case the changes are *reforms*;

3. transformations and modifications in which the actual or projected changes are such as to provoke at least the beginning of a substantial improvement in quality and equality—in this case the changes are *innovations*.

The first question to be asked is whether or not innovative processes can be imagined. My reply is affirmative. In the first chapter we examined some of the existing possibilities of such innovations and in the next chapter we will see others. There are events occurring in the world that have a recognizable innovative significance. Yet it is important to look beyond specific examples and to try to see in what way urban planning becomes an innovative process.

Someone may well ask, why urban planning? Quite frankly I do not have an adequate reply beyond a series of intuitions, which follow. To be concerned with how to modify innovatively that portion of sociophysical space called the city means, first of all, to enter within the domains of "competence" of all the citizens of the city. Schools, hospitals or even the banking system—certainly all of us are in one way or another affected by them. Often, however, these effects do not seem evident or we are not aware of them. Many of us also recognize that if these institutions were different, certainly life could be better. But many do not know what to do or do not feel competent to do it; people feel as if they are "incompetent." If we discuss, however, how to organize our way of life, our mode of living, very few people feel uninvolved or incompetent to say at least how they would like to live.

The second intuition about the importance of an innovative urban planning is the fact that to organize or reorganize urban sociophysical space means to interfere not with some but with many institutions, if not with all of them. This aspect is as important as the preceding one because there is no chance to succeed in transforming society without a simultaneous attack on a multiplicity of its organizational aspects.

The Humanization of Urban Space as a Revolutionary Process

In trying to examine what the concepts, methods, tools and, finally, the "products" of an alternative urban planning are, I will once again follow the indirect method, speaking first of what an alternative urban planning must *not* be.

In fact, this seems to me to be what emerged clearly enough in the last chapter, to which I will refer, implicitly or explicitly, as I continue. It is important to keep the present situation of urban planning in mind in order to have a clear view of the alternative city and society, the objective of this chapter. I do not, however, propose to design a new utopian and ideal "city of the sun." Rather, I hope to understand, and perhaps to help others understand, what points one should take up, what are the important issues in transforming urban planning and the city of today.

It is evident that if we were not in a modern, stratified capitalist society, modifications might be thinkable in terms of innovative processes in which masses of people, given clear ideas, could involve themselves. In our situation clear ideas, however, are not enough. This becomes evident when one examines why human beings initiated, and some will fight tooth and nail to preserve, those processes called institutions.

Institutions generate, regenerate, maintain and deepen the divisions of labor, the divisions of life, and the divisions that we refer to as social differentiations. The elimination of these states of affairs, to which we have attributed the key causes of the problems of society and the city, cannot take place today as the realization of a project to make institutions function better.

To aim at the disappearance of the divisions of labor, of life and of social differentiation means to undercut the chances of a number of men to control, dominate or exploit other men by entrenching themselves in and behind institutions, which in appearance and by general accord have a superhuman character. That clearly cannot happen except as the result of a process of radical transformations, of real innovations. The carrying out of a relatively broad project of reform is not enough. Institutions must be transformed.

I myself began with the conviction that the battle for the transformation of society should pass through a linked and articulated series of reforms of this or that institution, in this or that sector of economic life. I now find myself taking a different position, but it is not meant to be an attack on those with whom I was a companion in unsuccessful battles or those with whom I joined forces but failed to produce any great or even small leaps forward. What I propose, instead, is to convince them, too, of the impracticability of reform. Otherwise, a task such as the transformation of society will only be deferred to successive generations.

For this transformation it is necessary, even if it is not enough by itself, to have a number of clear concepts and objectives, and also to have clear a number of conditions in which firm refusals must be made. One runs a risk otherwise of doing just that which it is better to try to exorcise immediately—utopian planning. It would be a mistake for a number of reasons. Planning, envisioning, a utopia would give the mystifying impression that everything is clear in its final forms. This is not so and cannot be so, Campanella, B. F. Skinner, or Corbusier notwithstanding. And if a utopia were so designed one could be sure it would not be right.

The criteria and objectives for the future can be clear, but the fact remains that the objectives can take a variety of forms of sociophysical organization. There is no way for a single person or group of persons to imagine all of them or to choose the best one. This is even more obvious if one considers that the single person imagining specific utopian forms of organization nearly always ends up using his knowledge of what already exists and imagines the best that can possibly be derived from the best that he knows.

Finally, it is fundamental to try to be consistent; the organizational forms of a community must be intrinsic to its decisional processes. In other words, the same men who live those processes must produce the organizational forms, rather than the latter being predetermined and externally selected.

I would like to make it clear immediately that by this I do not mean that each and every moment of time all the citizens of every community must create or re-create everything from the beginning. Through knowledge and experience, though, they can become aware of and responsible for the rules, limitations and impositions through which they can achieve the organizational forms they have decided to give themselves.

I would like to give an example to explain what I mean when I say that we are not capable of making utopian designs. Let us suppose that we want to imagine an ideal community. We might establish a criterion or objective that in this community each member or group of members ought to have "rights to a house" and that each has an equal opportunity to satisfy this right.

A consistent form to be followed, given such a utopian design, is this. First of all, one may assume that the best solution to the problem can be had if the community (shorthand for all the citizens) controls the exercise of this right. At this point, then, two general directions can be taken. Housing might best be guaranteed if the community directly takes responsibility in supplying the houses. In other words, it is the community's task through a given organizational form to study people's needs for shelter and to assure that good planners and workers construct, for example, prefabricated houses that will satisfy all the basic biological and psychological needs of individual persons and families or even groups and categories of persons as house buyers or house users.

The second direction is that of those who believe in furnishing a series of more general rules that will guarantee a solution in harmony with specific preferences. Then, one provides people with individual pieces of land and basic building materials as well as with workers and specialists. In this way people can construct less standardized buildings more to their taste and organized inside as they prefer.

Both of these directions can be defined as utopian. They correspond to an ideal vision that is better than most situations that we know today. Both, however, are wrong ways of proceeding, as wrong as the present ways. First, one begins from the widespread but erroneous presupposition that there is a series, distinctive and distinguishable, of elementary biological and psychological needs that can be resolved separately, isolated from other social needs, which eventually can be totally satisfied by a process of addition.

It is equally wrong for present parameters to be used to imagine general solutions that would surely be sought and thought of in other forms if men come to be considered in more global terms. By that I mean considering people as members of communities of total people rather than as isolated individuals or consumers or as part of another partializing category. The needs for protection and for privacy that today are identified as primary needs of housing could tomorrow have various forms of satisfaction other than in institutionally shaped housing terms.

The weak point of modern writers who try to outline images of future society is deeply rooted in nearly universal models of atomized and sectoralized man, man as the sum of a finite number of needs, with institutions in their superhuman reality having the task of satisfying those needs. Total people either do not exist or are merely pawns of external forces.[13] Marcuse's analysis is supported: even in projecting ourselves into the future and trying to imagine a better future, we cannot but use those deeply ingrained, mind- and body-structuring models. Those utopian images become, then, those of a social organization where institutions function effectively and answer in an ideal manner that sum of needs that has become the central identity of man today.

Needs continue to be reified. They are, in other words, identified as and with objects that constitute in reality only the raw material for human experiences. But it is the experiences, and not "needs," that define man as man and distinguish him from animals whose absolute "needs" seem to be more directly satisfied by things or by much more specific, more simply biological-physiological experiences.

If we return to the list of requisites of the social quality of urban life, we see that none of them is a "need." They refer to human experiences and, as such, they identify the core of man as a social, experiencing being.[14] It is through those experiences that the total man, not divisible into various categories of needs, is defined.

To make the position crystal clear: Marxists as well as non-Marxists for the most part accept the model of man as a set of additive needs even if one or more are conceived of as being the most basic and central. While there are some exceptions, even a Marcuse in bemoaning modern society's "one-dimensional man" cannot shed himself of the distinctive need categories (the cultural and aesthetic, for example, as particularly underappreciated and trivialized in modern society). That is true even though he has glimpses of the importance of various human experiential categories that negate the original Marxist as well as non-Marxist Western philosophy's model of man as a set of separable needs.

Our problem, then, is to reconsider what it is that man in his totality has lost. We must reexamine those very parameters of the quality of urban life that cannot be obtained by improving institutional functioning. This is a vitally important objective because if radical changes in urban quality do not occur, the prospects of the urban future will be different from those forecast but only because they will be even worse than expected.

We must examine the contents and the objectives of radical changes in urban quality, rather than develop utopian forms or precise rules for their attainment.

Let us begin with a hypothetical situation, one in which a particular community decides to plan its future development and the organization of its social life rather than leave it to chance or to the working of present trends. Let us suppose that the community considers someone particularly useful for helping it. This person has already had experience in setting up and executing similar planning processes or programs of community development or else is skilled in a given sector. The community decides to ask for his help.

This starting point seems to be the description of a real situation. It seems to have all the ingredients normally found in a city at the start of an urban-planning process. There is the community, the urban planner (or other expert) and the plan to be drawn up.

This is a superficial and misleading impression, though. I have made at least three assumptions that are different from what actually occurs in reality. The first was my comment that the "community decides" to concern itself with its future development. By community I do not refer to a minority representation of citizens that decides for everyone but to all citizens or at least a substantial majority of them. That is very different from the situation in which the great bulk of citizens not only has not expressed its will but does not even know that a plan is being born.

The second assumption concerns the meaning of planning or programming future development, of organizing the community's social life. This does not mean establishing land uses for a given and prefixed number of years, as one generally does through a traditional urban plan—as we have seen in the preceding chapter.

The third assumption concerns the role of the urban planner (or expert). To suggest, as I have, that he is employed to help the community relates to the first assumption. The "help" that is given in traditional urban planning when he becomes the community's so-called interpreter and makes its decisions occurs only within a narrow circle of people who have relevant expert qualifications and/or have the power to perform such decisively interpretive roles.

We might ask ourselves if these distinctions are basic. If we look at real cities and towns, the response must be affirmative. In fact, these are the aspects that serve to identify the people who shape the plan and who define its very essence. In a word, the fact that in the hypothetical situation the whole community is participating in the planning process right from the start and over the period of time makes it not only a different but also an *alternative* situation.

Some may assert that, in reality, the opening of the processes of the plan to include people is being attempted but people do not respond. In reply to such an argument I will only reiterate what I have said before: the type of participation attempted so far could have only those limited results. This was evident in the results of the studies focusing specifically on this matter of citizen participation which we analyzed in the first chapter.

The paucity of results of these attempts at greater participation, then, does not bear on and cannot modify the basic hypothesis. That hypothesis is that present urban society cannot be transformed unless one can intervene substantially and radically; that is, at the nodal points implied by the three aforementioned differences between the contemporary reality in urban planning and a more innovative process.

Let us suppose that there are municipal administrations convinced of everything suggested so far in regard to the present situation's being an extremely negative one. Further, suppose they agree that the people themselves must participate in the processes of organizing their lives in the community. This situation is not entirely fictitious. In fact, by now many municipal administrations exist whose legislative and ideological approaches, at least verbally, are explicitly of this type.

Such local administrations are certainly aware that one does not overcome apathy by carrying out an abstract "antiapathy" campaign. On the other hand, they understand that a radical transformation in the processes of organizing the life of the community cannot be effected—as long as one accepts "administrators" or "representatives." They use laws that already are inadequate or even distorted in their very mechanisms (as we have seen in traditional planning procedures). Again there can be no such transformation if one accepts as the citizens' representatives—their spokesmen and the decision-making substitutes—local administrations who do not, moreover, have financial means to satisfy even those few needs of people that today are officially recognized. Nor can there be a transformation if citizens are to be the represented and only have power every now and then to vote for the few to represent them.

If anyone really believes in citizen participation, he should not, moreover, settle for accepting or being content with procedures wherein those represented may only make suggestions. Even local administrations desirous of citizen participation provide the representatives with their experts to make the decisions, and in the name of the citizenry.

If there really are administrators who are aware of all this, they will begin to understand that alternative planning processes can begin when they, the administrators, start to formulate proposals and programs framed in far less specialized terms than are usual. Such administrators and officials need also to understand that they themselves must become less specialized, less the experts in representing or deciding for others.[15]

What does this mean concretely? First of all, it means that even in current circumstances it is not useful to decide at the city-council level to make a plan. Of course there are state laws to be respected, giving jurisdiction to city councils, as well as good reasons for councils to act to keep speculative development and redevelopment under control. But discussing with the community from the earliest moments the reasons why a plan seems to be necessary, is one way the traditional plan can be broadened and opened up. Such discussion also signifies that one now considers the "experts" involved to be the assistants to the real producers of the plan: the citizens. The reader's major reaction to this may be a feeling that we are proposing the substitution of an unworkable, romantic plan of direct democracy for the tried and true representative democracy.

In any event, that change in roles on the part of local officials and experts, however difficult politically and psychologically, is the second step if one wants to have a planning process that has as one of its primary objectives that of maximizing the presence of citizens therein. Such a goal is impossible if, as presently happens, efforts are directed toward having the plan made quickly and achieving its aims rapidly by obtaining necessary higher-level governmental or judicial approval without troubles.

If we begin to realize such changes, we have begun an innovative process whose next step, instead of collecting the traditional statistical data, will be to begin discussion with a certain number of citizens on the present conditions of their life in the city and their view of their problems there. The images that emerge from such discussion can then become the object of more discussion that leads to judgments, evaluations and ideas from an increasingly larger number of citizens. In that manner not only the past and present but also the future can become objects of exploration. Various alternative futures and their modes of attainment can become the public business of the many rather than of the very few.

The fundamental points of this kind of process as sketched so far are these:

1. knowledge is generated with the participation of everybody (and not only knowledge, as we will see);

2. the process is continued through time with periodic reexaminations, not merely an initial examination;

3. the examination focuses, from the beginning, on matters and concepts usually reserved for experts, not only on so-called personal or private experiences.

One of the very first of the concepts to be examined critically is the concept of the city as a place characterized by certain physical aspects and factors to which are then added social ones. Another conceptual myth to be destroyed is that which says the physical characteristics of urban space exist per se and can thus be studied separately and that prescriptions that concern them can be given separately, apart from the human beings who have affected these aspects of urban space and continue to do so by their presence.

Such critical conceptual reexamination, though, does not comprise a separate, specialized operation to be entrusted only to particular people of presumed special competence in presumed "intellectual" or "professional" operations. It is precisely when persons of various ages and conditions bring to such reconsideration the meanings of their various experiences that the important sociophysical space emerges almost inevitably from the otherwise socially meaningless physical space. Consequently, it no longer emerges only with aesthetic and functional associations of the specialists, as presently happens, but it is inextricably linked to and largely defined by the human events that have occurred, do occur and may occur in it. The city and its many sociophysical spaces thus emerge from the rich human experience of the past and from the multiplicity of variations therein which the city has provoked and continues to provoke.

The complex image or images that emerge will perhaps not be accepted by all and will not always or even for the most part be uniform; they may often be discordant. But such a "pluralism" has a cementing uniformity in the shared right/duty of participating in civic self-management.

In any case, it is clear that a many-faceted image, with various question marks or even conflicts to be clarified in successive phases is more important and more interesting than the current one. Certainly, it is not the clearly structured, unequivocal and stereotyped, although incomplete, picture of a map of the city and its parts that the urban planner traditionally constructs with his methods and data.

Another necessity in such a process is to regard each aspect or facet of the community as having its particular dynamic. While that concept is given lip service by urban planners, their operations violate it and treat the city as a set of often fixed structures. Nothing is entirely static, however. Some facets change slowly while others change rapidly, but all change. These are aspects with which each of us experiments in everyday life: within the global meaning of the total experience as a human being a person sees himself aging as well as his automobile and his house. The comparatively slow or seemingly slow rhythms of change in buildings or other physical structures permit the urban planner to treat as fixed, matters that popular participation would demonstrate are most dynamic aspects of the sociophysical.

In the light of this proposal that envisions the beginning of the plan as a collective event involving the entire community, surely someone will ask: what role does the technician/urban planner have in all this? And, can one think about such a planning process without first thinking about taking land into public ownership? Further, how is it that I can say this process can begin when the citizen participation that is essential is precisely one of present planning's fundamental deficiencies?

As a reply to the first question, I must say that not only the urban planner but also the administrators have important roles in this process. The urban planner, the expert (and in a manner not very different, the administrator), has a double task in the initial phase. First, he must make his fund of existing knowledge and experience available to the citizens. Second, and even more importantly, he must open up and involve himself in the process of demystifying this knowledge and his own image as "expert." Also, in successive phases he may have the important role of being a coordinator, having had experiences with various communities and being in a position to be of such assistance, even if no longer having the role of being the exclusive interpreter of the community's needs regarding space.

One of the simple but important things he might do is "translate" his professional terminology into everyday language so that it will be accessible to most people. Such an initiative would contrast sharply with that of the expert who entrenches himself behind his jargon as if seeking to consolidate his status by generating amazed and astonished admiration through difficult expressions meant only for the few who are "competent" to understand.[16]

As for the problem of public ownership of land, I agree that it is fundamental but it should not be treated as the priority. Or, rather, I do not believe that one will succeed in resolving the matter satisfactorily if land uses and zoning continue to be treated as they are today, both in terms of decisional process as well as substance.

We often overlook the fact that speculation in building has not been the work primarily of dishonest entrepreneurs or contractors. It has occurred with the cooperation of hundreds of administrators (whether fully aware of the consequences or not), and to the disadvantage of thousands upon thousands of people. But we should not forget that thousands upon thousands of people from every social class have agreed, without the least hesitation, to rent and buy shoddy houses and buildings built by speculative processes, often in areas far better suited for other uses. This is not an accusation; it is merely a statement of fact.

But what if, instead, the community or at least the bulk of people were convinced that such modes of using land prevent qualitatively better development of the sociophysical environment? In such an event, I am convinced that we would engage in political actions to obtain new laws and measures concerning property and land use with much greater chances of success than we have had in all these years in the United States as well as in Italy. In fact, until now these actions have typically been initiated by small intellectual elites. Then part of the political class joined in, usually in an indecisive way because of fears of being misunderstood. They faced a citizenry uneducated in matters of land use or urban planning generally and often manipulatable by demogogic cries about subverting the free-enterprise system. Then the "grass roots" followed but, indeed, they only understood part of what was involved and often only partly agreed.

On the other hand, how could and how can the majority of the population be convinced of the value and necessity of a political fight for the public ownership or use of land? How can it be convinced, especially when the so-called public power has hardly ever provided examples of how land might be used in a more socially and humanly beneficial way than private speculators have used it? With regard to this aspect, then, the real priority is to understand and make it understood that the "urbanization" of space, especially in the modern epoch when physical space has become a commodity of such economic exchange value, has not meant, as it should have meant, investing this space with human values, to "humanize" it. And that is true for public as well as private developments.

On the contrary, the physical forms of modern settlement have become the realization as well as the catalyst of the spatial disaggregation of society and of the dehumanization of physical space.

It is important to understand and moreover, make it understood, that there *is* an alternative. Urban space can be humanized; urbanization can become humanization if everyone may contribute, especially the large numbers of people who are presently subject to the discrimination and alienation imposed by the institutionalization of society.

Returning now to the third question posed above, how is it possible to believe that the participation that I have described as essential for these initiatives can be achieved? Doesn't the lack of participation immediately indicate that such a route is impracticable, unrealistic and utopian? In replying, I would like to refer to the conclustions of the empirical studies which I examined in Chapter One. That research in communities in several countries, countries which differ from each other in terms of historical past and current political situation, showed that the citizenries, even when not participating in various kinds of community affairs, are nevertheless interested in many of them. The greater part of participation passivity, that is, mere interest without more active involvement, comes from frustration, from citizens' inability to believe that they can in any manner influence what occurs in the community outside of the smallest, most privatized sociophysical spaces.

Now, the process I have proposed touches exactly this matter, therefore making it plausible to count on a decided increase in active participation. What actually takes place in the process? First of all, once this process of community self-knowledge is begun, it does not happen once and for all but continues and updates itself through the passage of time. Of course, this needs organization but it does not need to be institutionalized. In fact, contrary to the criteria defining an institution, such organization needs to be inclusive, comprehensive, integrative and open. That is because it is based on the principle that everyone can take part in every phase of the process. Rules establishing who does and who does not have a right in this regard consequently are not necessary. Particular qualifications or requirements for knowing certain secret or confidential information, necessary in the case of institutions, are not necessary here.

I repeat, though, that various aspects of organization—specific calendar dates for courses of action and persons who assume particular tasks—are certainly necessary. Those kinds of things are possible without hierarchical structures by following simple patterns of rotation that impede the development of organizational ladders.

Someone may object that in any case such a design cannot be realized. Even if participation is generalized—if, for example, a doctor participates along with more ordinary people—someone easily and quickly assumes the dominant role, in this case the doctor, having high social status and being more experienced and skilled in speaking than most people.

This will tend to be the case, especially in the beginning. The reason is that we are starting from a very particular situation. As soon as a doctor begins to speak, he is given more attention than a layman who has no such title. Apart from how he says what he says, his proposals are at least examined more attentively than those of others. But if this is the situation from which we start today, we do not want it to be, nor need it be, the one tomorrow. Beginning from what happens today, the first step is to make it understood that the only qualification for participating in such urban planning processes is that of being a human being.

Obviously, in any human organization there are persons who exert more influence and others who are more easily influenced. One cannot imagine this to be otherwise. But the distinction between this type of influence and that actually exercised today is that the latter in most instances is influence that does not derive from a person's human qualities. It derives rather from the prestige of one's titles, one's positions and one's social class.[17]

The alternative urban planning process that I have started to outline has another important characteristic. Distinctive stages in the planning process are not envisaged: there are not and cannot be separate phases of gathering knowledge, of decision making and then of execution. This is also true with regard to the traditional planning process. But because that process is mystified by encrusted although fictitious divisions with their various signs and symbols, it is not thought to be true. Indeed, it is said that there is a first phase when the expert collects his data, followed by a second, data-analysis phase, followed by his suggested decisions or the decisional alternatives. In the next phase decisions are taken by the community representatives, and then comes the phase of decisional execution and, here, finally, everyone may intervene.[18]

It is not like this, as we have seen. The plan is traditionally a unit even if there seem to be distinctive forms in which it is studied, prepared and realized. It comes into, and remains in, the hands of the few from the very beginning through its implementation. The urban planner and his commanders, sharing the concepts and logic that we have been criticizing, are those few present, and consistently so, in, under, or behind every apparently distinctive stage.

In our alternative process these phases are integrated and stress is given to just this unitary aspect. Decision-making and the actions that follow are opened to all precisely when and if participation in the "production" and "preproduction" operations of initial data collection and knowledge generation are also opened to all.

In fact, it is true that what and how something is studied conditions or at least turns one toward certain measures and actions rather than others. Traditional urban planning and its procedures are an example of this, as we have seen. In the process I propose, the conceptual categories of gaining knowledge or of data-gathering and analysis, of deciding or of making decisions, and of executing or administering them are transcended. But there is something even more important to realize. Many of the events labeled data collection or generating information or knowledge are actually important participant decisional actions. And they are actions in which people other than experts have vital contributions to make.

For example, what if elderly persons were asked to speak about their problems, about past events in the city, and were given the opportunity, as a matter of right, to speak about the possibilities of the future? Why should not such important actions be regarded as "aid to the elderly"? It would surely result in making them feel more alive and present and a part of the community than merely guaranteeing their physical survival, as the food of the astronauts would do. And it would not be a transient feeling or sensation; the fact of their becoming more a part of the community, more present in the community, would be a reality affecting the whole community.

And are not those initiatives that stimulate people to learn, to know and to evaluate actions as we are proposing also actions of "adult education"? These surely are processes of integral education, of political education.

A third and very important result of the alternative process proposed is the following: once the informational and decisional processes are opened, without any limitations or secrets, to the entire community, a single plan for the use of physical space is no longer possible. There must instead be a series of actions, programs, or projects whose forecasts and directions cannot be represented in a model such as the master plan.[19]

As we have seen, a master plan consists of a two-dimensional and static representation of functions to be located spatially in a certain future period (the maps of the plan). The plan also consists of a series of rules, norms, or codes dealing with two aspects. The first organizes what ought to happen in the third, physical dimension not represented by the maps (that is, rules establishing height of buildings, distances between them, their appearance, and so on). The other aspect concerns the restrictions and procedures and duties that must be respected or followed by builders and developers of the physical structures and infrastructures.

All this, however, as we have already remarked, does not really allow for foresight into how the forms of social life might be organized in physical space. On the contrary, everything specified in these plans, such as the decomposition of functions or the creation of functionally homogeneous areas, works exactly against the creation of urbanized space where man might live in a less sectoralized and more undifferentiated way. The social "quality" of what is put into those plans (by social quality I include aesthetic quality) is left to those who actualize the plan. Whether or not they are speculative builders or developers in the usual sense of the term, they almost always apply only criteria of an economic nature.

Once the entire decisional process really rests in the hands of the citizenry, the potential actions can be discussed and examined in the light of broader and more significant criteria than, for example, what functions are most suitable for a given area or how many square yards are allowed for a particular construction.

By this I do not mean to say that everyone decides about everything. Everyone is not interested in everything. Nor can they be. I mean that everyone *can* participate. Nor do I mean that evaluative criteria should be merely improvised; that each time they should be accidentally different. It is clearly necessary to create instruments that, if they are to be more sophisticated, must not therefore be more difficult to understand and use. In other words, they must not belong more to the specialists. Nor need they. The modern equation of advanced technology with technocrats, of sophistication with esoteric expertise, is wrong. It is a question of creating, adapting and using instruments and criteria, I repeat once more, that take into account the life and total experience of the human being and the fact that precisely because of his human nature man's experiences are those of a dynamic system. And most people, for the most part, choose to live in those domains where they are permitted to be at least relatively successful dynamic systems. This is not the case when people are treated as static components or cogs or fixed and frozen entities in larger mechanical or even biological kinds of systems.

This does not mean, of course, that everything in the community must therefore be considered in need of change. It is in the community, in the lives of its members, that values and aspects of community life that are most apt to be evaluated as worth conserving are to be found, only subject to slowly varying changes.

A community is not, however, a closed and isolated world. What occurs in it is influenced by and influences what occurs in others. In this regard, it may be noted that so far I have not spoken about "decisional levels," of decisions at the community level and higher levels. In fact, my basic assumption is that we should aim at a future in which only one level of power exists, in which various communities, that is, citizenries, interact as equals (putting aside the matter of community size).

Is this possible? This demands a greater exploration than space permits. But I am convinced that it is possible when I undertake the following mental exercise. Let us take a nation such as Italy. Why is decision-making necessary on the national, regional and municipal levels? Suppose that, given modern means of transmitting communication and processing information, we consider Italy to be not a "nation" but simply a large city subdivided into a certain number of sub-communities. And to each of these is entrusted the stimulating and important task of dealing dialectically with its own problems together with the overall problems of the city. We would then appreciate that so-called levels of decision-making are aspects of a single encompassing system of decision-making. And creating distinctive "tiers" of government is not the only alternative to many, small jurisdictions in an overall urban anarchy.

So, too, are national, regional (provincial, state and others) and local levels aspects of a single system of decisional processes. Removing the distinctive levels does pose the grand problems of the interchange of information among many people over very large areas. Those, however, are merely technical problems that present computer technology can solve, as Beer has shown.[20] To accomplish that, it is necessary to realize a process having the connotation of "decentralization." It is to commit oneself to the values of citizen participation as overriding such considerations as the high economic costs of certain large-scale facilities and services and opting for a truly unilevel decentralized system.

The key problem of decentralization is not that of decisional feasibility among masses of people nor of choosing the appropriate size of the basic unit of decision-making. In fact, I do not believe that abstract or general answers can be given to the latter matter without falling into the same kind of useless debate that every now and then flares up in urban planning in discussions about the optimal dimensions of the city. The technically appropriate decisional devices for mass democracy can easily be put in place.

In order to overcome the vertical, segmented kind of organization of contemporary society and to arrive at a predominantly, if not totally, horizontal organization, the process must start from the actual situations, experiences and characteristics of the various socio-urban organizations of today. It must also conceive of the boundaries of the community not as fixed but as subjective and subject to changes, without predetermining the "optimal dimensions." Those will emerge as the process proceeds.

Certainly, what Schumacher proposes seems intuitively correct: we should aim at the "small dimension" whenever possible. That fits the idea, one that seems full of possibilities, of having a village kind of organization even within the most extensive urban aggregations.[21]

It is of basic importance, however, that the organizational model respect and rest on the requirements described earlier. That is, one should try to establish that dimension of the community that can facilitate, as urbanization continues, the abolition of the divisions of labor and life. It is this dimension, therefore, that can contribute to the quality and equality of human life in a space not simply urbanized but also "humanized."

This means that the dimensions of the community must be such as to allow for the process of its members' self-knowledge and decision-making with respect to their own forms of life. For the community it is a globalizing process/project that requires concern for people in other communities and in the world community, and also for so-called economic decisions, and first among these, decisions on the forms of industrial development. This does not mean that if we want to guarantee industrial development, we have to have the large dimension—as we now do in modern society. Such questions as the scale and character of industry must be open for consideration by all members of communities and by confederations or other associations of communities. These are questions that are considered today only by the few directly involved and by regulatory officials of government and generally on the basis of economic rather than other dimensions of appropriate uses of resources and of urban sociophysical space.

In fact, it is not only Schumacher and other scholars and specialists who argue persuasively for an alternative. In some Third World countries there are already experiences and applications of it. That is, both theory and practice now demonstrate how the dimensions of industrial production can be reduced. That is true even for some products earlier considered only suitable for large, concentrated and centralized complexes operating according to assembly-line processes.

Returning, then, to the preceding proposal, there is little doubt about people being interested and participating in a truly opened planning process. There is little doubt, that is, if, and only if, we open up urban planning processes in such a way as to make them really become total processes of organizing the forms of a community life in which all can participate and in which all can decide. There is such little doubt that many of those who claim people are not interested in participating are adamantly opposed to the opening because they decry the mass participation that would occur.

By beginning with the community and by putting the accent on the total person, we have chosen the surest way, and perhaps the only workable way, to create an alternative to the contemporary forms of societal organization. This is the only possible way to begin to heal the cuts, to close the gaps, that characterize everyday life and to encapsulate the decisional processes that today are separated into a multitude of falsely divided processes and mystified as superhuman under the labels of institutions. By so proceeding, we can participate in an evolutionary revolution, that is, one not requiring blood and violence. The criteria of and objectives for the process of transformation can be specified, although not its final forms.

For a Strategy of Transformation

I have dwelt so far on the contents and objectives of an alternative process of planning using criteria that were previously defined as innovative, which term signifies fundamental change. In fact, we have seen that this alternative process of planning is formed in such a way as to solicit, facilitate and stimulate transformations in the positive senses of quality and equality in urban social life. The process does this by substantially modifying the workings of some of the factors we have seen as crucial to these transformations, such as interest in participation and self-confidence on the one hand and access to information and decision-making on the other.

Still, some major objections to these proposals and conclusions can be raised.

1. Some might hold that my description of urban planning, its practice and its consequences is persuasive; that as a specialized discipline it probably should be eliminated and at once. This, however, does not imply that one must do battle with the other fields of specialization on which progress has depended and depends. Besides, such a person might go on to say, if certain specialists seem superfluous at the practical level, others most certainly do not. Think of the engineer in charge of constructing bridges and skyscrapers, but even more, think of certain specialists in the sector of medicine. Would we be able to, or would we like to, renounce the work and ability of the surgeon specialized in a very delicate kind of operation?

2. A contrary objection that might be raised by someone who considers specialization (a manifestation of the division of labor) to be totally negative is that, actually, the proposal to open up urban

planning does not eliminate other specialists. In other words, there is no chain reaction leading to more complete revolutionary modifications in society.

3. A variation on that objection is the position that revolutionary changes can only be provoked by modifying the "economic system," the "productive system." Indeed, such a "left" or radical critic might maintain that capitalist society is made in such a way that any transformation whatsoever, even a radical transformation of the "urban system," is rapidly isolated, encapsulated, inserted into and digested by the system. In the present state such a transformation can even be a useful and unexpected help to the system as it copes with annoying if not deeply disturbing urban problems. Such a critic might add that the subtracting of energies from the fundamental battle against the economic system and from the "class struggle" is decidedly against the interests of the working class. Once the latter has won power and has substituted its own representatives for the present managers of power, it will be able to transform such urban disciplines as urban planning and architecture by modifying the demands of its users.[22]

Beginning with the matter of specialization, I would like to examine these objections precisely in order to clear the ground and to see where and how much they touch upon what has been said so far.

It is evident that the creation of increasingly higher levels of specialization has not been only a means of development of the disciplines themselves. The specialist acquires a social status that distinguishes him from the nonspecialist and from the lesser specialists at every level. Some say that specialization has always existed. There were always people who were agricultural workers or artisans who only worked at a specific craft. In a similar way, there have always been lines of demarcation between those who did work requiring primarily physical strength and those, instead, who did intellectual work and also between those who carried out orders and those who gave them.

Even if this was true in the past, however, it must be said that only in the industrial epoch has specialization entailed such an atomization of actions and division of partialized roles.[23] Tied to the creation of levels of specialization, which also give an identity to persons according to such partialized roles, is the fragmentation of the same social classes through which the dominant groups more easily maintain control.

Precisely in reference to the "necessity" and "indispensability" of some groups of specialists, then, the situation is much less black and white than it seems at first sight. First of all, it seems to me that nearly everyone accepts the fact that many fields of specialization can be eliminated.[24] With careful investigation I am convinced that we would reach the conclusion that they could be carried on with equally good results by nonspecialized persons and in organizations where different people performed on the basis of principles of rotation. (In many cases this could bring even better results than those presently obtained through specialists or professionals protected by the power of organizations and corporate conspiracies of silence.)

There is the usual objection, however, that if this were brought about, the organization of society would have to pay very high costs. One need only consider, for example, the preparation required to train the sufficiently large number of skilled surgeons, now specializing in one or another kind of surgery, needed to make a rotational procedure effective. It is also said that this specialty, like other activities, requires an interest and a natural ability that few have. I agree.

My principal objectives are to eliminate the "status" of the specialist and his privileges and to do away with his being partialized and treating others as partialized. In the case of medicine it would be far better to see doctors treating people as human beings rather than as "patients." In other words, specialization may exist without specialists. Being involved in a series of rotational activities of various kinds may impede the formation of categories of people whose mentalities and basic identities are those of narrow specialists. Moreover, considering that there are a certain number of manual and routine activities in human organization that must continue, at least for a while, the egalitarian principle of everyone's sharing in such tasks might well be adopted.

As for the economic cost of eliminating specialists, or even only certain ones, it seems that it might be much less than is suspected. All the privileges that so many specialists now enjoy might make an increase in their number, accompanied by a "normalization" of their status, economical in general terms.

On the other hand, precisely on the level of the disciplines themselves, sectoralization and specialization are producing such a degree of partialization that the development of some of these superspecialized sectors seems sometimes to be quite distorted. Some have lost any reference whatsoever to their general aim of serving man and promoting his human welfare.

The development of systems thinking whose philosophic base aims at overcoming a one-sided vision of the world as sectoralized, separated and divided into specialized parts suggests that the argument in support of specialization for more advanced intellectual developments is not valid. In fact, we must think increasingly in opposite terms, of wholes, of totalities, of total systems and persons.

In any event we must examine also the type of objections summarized in points "2" and "3" above. They include certain aspects, touched upon earlier, that have been at the center of debate in European as well as American Marxist circles for some time now.

The most traditional Marxist interpretations focus on the economic aspects, on the forms and ownership of the means of production, as the principal objective to attack for the transformation of society. Other Marxists, however, see urban reality as the major modifying factor of the relations of production, even if at the present stage it is still not a transforming factor. For Lefebvre, "space and the politics of space express social relations, but [also] react on them."[25]

The argument as to whether one must start with the economic revolution in order to arrive at the urban revolution, or vice versa, seems to me to have in distinguishing the two terms a wrong starting point. First of all, productive and urban organizations are governed by the same general rules and determined by the same logic; rules and a logic that, in turn, are expressions of a given overall system. In fact, urban man has been able to be alienated, separated and partialized from the moment in which this duality was produced in his place of work and in society. It was substantiated precisely by the forced division between the workplace and life-space as parts of two separate worlds and systems. Thus, by carrying forward alternative processes in the organization of urban social life, one touches upon, transforms and revolutionizes the entire system because the barriers protecting the productive sector are not respected.

The man who recaptures the totality of his humanity and demands that he be considered in total terms, that is, more than in terms of physiological needs, more than in separate parts or pieces, must also refuse to accept the separation between economic and noneconomic aspects.[26]

Very often this kind of criticism that stresses the primacy of the economic battlefield is expressed precisely by those on the left who have not succeeded in getting beyond the idea that the best organization of urban social life consists of organizing housing (public housing) and services (public services).

Lefebvre can be counterposed to that position and we cite him again: "a vast program that would also be a project of transforming everyday life, that would no longer have any relation either with a repressive and banal form of urban planning or with a constrictive organization of territory, this is the first political truth that must penetrate what remains of the French left in order to renew it."[27]

In order to create a process of alternative urban organization, one must open to everyone the knowledge and the right to participate in making decisions with a strategy of progressively broadening participation therein. This cannot be left to chance, considering that one must begin from present conditions.

There is, indeed, the risk that "the system" will encapsulate this attempt at such an alternative, since it acts and reacts in opposition to every innovative change by reducing each to the level of a reformist adjustment when not actually turning changes to "its" advantage. There is also the danger that the dominated classes may reject the content of these innovations.

But the risk with the "revolution" in the traditional sense is another. That revolution, to abolish private ownership of the means of production, demonstrated its inadequacy. Having substituted for one class in power another made up of the representatives of the formerly dominated classes, it does not succeed in freeing itself either from the specialists who act as representatives or from specialists of the other institutions.

Nor can I agree with Lefebvre that the architect, the urban planner, in short, the urban specialist, can still be "the condenser" of various new relations being constructed even if they are no longer of a capitalist character. I disagree because the urban specialist is not a neutral technician with neutral tools. His own role as specialist cannot survive a truly innovative transformation of relations in the city.

Lefebvre himself poses urban self-management as his primary goal and sees in it the potential for bringing about self-management in the industrial sector as well.[28] It seems to me, though, that urban self-management must not stop with the organization of the city nor even of its so-called everyday life. It must become extended throughout the urban society. This can only occur by inducing, and being induced by, a progressive de-alienation of the human being based on recapturing or re-creating his vitality and self-confidence. This precludes specialization *as it is presently structured* and every suprahuman function for persons, roles or institutions. And it precludes the practice by the urban planner of his traditionally structured specialty.

5

AN EXPERIENCE OF PARTICIPATION IN URBAN PLANNING

The Plan for the Historical Center of Faenza

A few introductory comments are necessary to make the significance of the Faenza experience, which will be examined in detail, understandable. Faenza, located in the region of Emilia-Romagna in northern Italy, is a small city with some 54,000 inhabitants and a notable historical tradition. Several centuries ago it was the most famous center of Europe for the production of ceramics.[1] A few years ago the regional government of Emilia Romagna, which has general jurisdiction over urban planning matters, ordered the Faenza municipal government to draw up a detailed plan for its historical center.

After World War II, in Faenza as in many Italian cities the inhabitants of the oldest section, the historical center, who earned average or high incomes and wanted new housing left for the expanding suburbs. The center was left with areas in which the most disinherited and marginalized classes live in increasingly miserable conditions while in other areas speculative development occurred in which space was used according to its profitability. Those areas were devoted to commercial and business activities and in some cases, given the high artistic-historical value of some places, residences for the privileged who were able to meet the comparatively very high costs of living in restored palaces or other monumental buildings.

Although differences exist, the situation of Italian historical centers recalls that of the oldest parts of American cities, the so-called core or downtown areas of inner cities.[2] There is a tendency with respect to the American city to characterize certain areas according to their most obvious social and economic problems, for example, racial ghettoes, and so-called gray areas. There is still a tendency to think that these constitute special problems within a healthy urban environment although gradually it has come to be understood that the pattern is different. Not only some areas but the entire American inner city (in Italian terms the historical center) is decaying generally. In fact, the present urban crisis, and especially the fiscal crisis of the American city, has its source in the overall decadence of urban areas, indeed, of urban society itself.[3]

But let us return to the similar although less severe problems of Faenza. The starting point of the experience to be described here was the scope of the mandate that the municipal administration gave for a plan to revitalize the historical center. The planners were to take into account not only problems of the buildings such as various palaces of historical value that were empty, some near collapse, but also social decadence. In other words, the plan also had to deal with a considerable part of the city's elderly population as well as those in precarious economic conditions and those living in very poor dwellings (often but not always being the same people).

The original commission mentioned that the plan should be produced through the participation of the citizens.[4] This fact, however, deserves a small comment. What was intended was nothing more than or different from what is traditionally known as a participatory urban planning process: to invite people to some public meetings to discuss or to hear about some phases of the plan's formation.

Another fact deserves attention here. What exactly the municipal administration meant by the goal of "revitalization" of the historical center was not made clear in the commission nor in subsequent discussions. The only clear intention appeared to be that of anticipating the building of some public housing in the historical center, a relatively new policy in Italian cities. Previously, public housing had been located in the outlying areas rather than in the center of the city.[5]

The final important background factor to be mentioned here is that work on the plan began some months before municipal elections. The administration in office at the time was a so-called center-left administration, formed by representatives of the Christian Democratic, the Republican and Socialist parties.[6] The elections caused a significant change in Faenza by bringing a representative of the Communist Party to the office of mayor while the executive committee of the city council was now made up of Socialists and Communists.

Work on the plan had started to become innovative before that election. The group of four professional urban planners that was originally commissioned, of which I was one, was asked to prepare a program on the basis of which the plan would then be drawn up.[7] Following the usual procedures, this group began to gather the traditional data in the forms described in the third chapter. On the basis of such data, then, the first outline was organized more or less around these subjects:

- the role of the historical center in the county and in the region;
- housing and its role in the historical center;
- industrial-commercial-business activities and their distribution in the region;
- the problems of urban traffic, especially in relation to the need for parking places and parking lots;
- the working schedule for drawing up the plan.

It was within such a traditional program that the first innovative element appeared. In preparing the plan, those involved decided to begin with an extensive social investigation of a random sample of 800 persons living not only in the historical center but throughout the county.[8]

Terming this an innovative element not only derives from the fact that usually even the most famous plans do not anticipate a social investigation at their base or, if they do, such an investigation is very restricted and often carried out after the plan is already drawn up.[9] The innovative aspect also lies in the approach of the investigation and in its intended use.

From the time of the drawing up of the plan's work program, in fact, it was clear that the investigation should not constitute only a source of knowledge for the urban planners involved but should also be an effort at self-knowledge for these very same citizens in the sample as well as for others. That required a major feedback stage at the very least. Further, the investigation was also identified with the initiation of a strategy to involve larger strata of the population in a process of demystifying urban planning and involving them in a process of self-planning. This was begun with those who were part of the random sampling.

In substance, an alternative route modifying nodal aspects of traditional planning procedures was proposed affecting first, those who act in the traditional formation of a plan; second, the methods, instruments and models shaping the planning action and, third, the contents of the plan itself.

Let us take a detailed look at the traditional sequence of events and then at the alternatives developed in Faenza. The first change concerned those who act in the formational phase of the plan. Books and articles dealing with urban planning speak of various phases: selecting objectives, collecting data, preparing proposals, discussion and approval of proposals and the management of the plan of its execution. Traditionally, three categories of acting subjects in such phases are singled out: the local officials who, as representatives of the community, define and establish the objectives, then adopt the plan and, finally, manage the realization of it; the urban planning experts whose role is to indicate the best technical solutions for achieving the objectives proposed and, then, the people who participate in the discussions on the proposals of the technicians, thus corroborating the plan's democratic dimensions.

As we have already seen, these assumptions are not valid. In particular the public meeting is certainly not the time or place for discussing or making fundamental choices. A large and representative or cross-sectional group, like the one selected for the sample survey, has the advantage of containing far more people than those few who are already practiced in participating in public meetings. It allows all the advocates (officials, experts and the people) to be present from the beginning in the planning process whereas usually only the first two are present.

The second change concerned the methods, models and instruments shaping the planning action. Basically, one should no longer separate those who carry out the analysis, those who compile the data, from those who are carriers, the bearers, of the problems to be analyzed—that is, those who become the numbers of the abstract "data." The concept of participatory research, of action-research, already used in certain North American situations, comes close to this basis of the Faenza experience.[10]

The presence of those who are experiencing the problems allows for the overcoming of methods and instruments based on schematic and fragmented analytical categories, which in turn contribute to the fragmentation and schematization of the plan's proposals.

The third area of change was in the contents of the plan. We have seen that both administrators and technicians (urban planners) have accepted, and continue to propose, a vision of urban life according to artificially separated categories on which the organization of industrial society is based. A high human quality of urban life and equal opportunity of access to such quality cannot be generated from the unidimensional values and partialized models of people and of their needs and desires now used in planning processes in the United States as in Italy.

A participatory plan is one prepared from the very beginning with the participation of people in the totality of their sociophysical-economic-political-cultural moments, moments that are unified although multifaceted. Such a plan becomes the expression of man's life as a social being in his everyday experience. We stress the point that the data-gathering operation of the social research was far more than mere data-gathering and far more than mere social research. Through it there was the possibility of having citizens present who were real persons with real characteristics rather than abstract categories or statistical artifacts. Those characteristics contrast sharply with the traditional census and population data representing very few and fragmented characteristics of the members of the community. By comparing the results of the first phase of work at Faenza with successive events, it has become clear *a posteriori* that the social investigation constitutes the fundamental element for modifying urban planning methods, the cornerstone of an alternative, truly participatory process.

As noted above, as a first step the experts had prepared a document containing their proposals for further work. Meanwhile the new administration of the left, having taken office, organized a number of public meetings of "citizen participation" to discuss the document and proposals. This was a first symbolic act of the new municipal management "of the people."

Such meetings were, however, convoked in the usual way, that is, through public posters and letters to the representatives of business, cultural and other associations, labor unions, and to the neighborhood or district councils (*consigli di quartiere*).[11] In these public assemblies restricted discussions and debates involved representatives of various groups and interests as well as the technicians. The latter, as usual, paused to discourse abstractly on such abstruse matters as the "role of the historical center." No one asked himself (at least aloud) if this figure of speech (termed "metonymy") which has become urban-planning jargon corresponds with human reality in any shape, manner or form.[12]

In other words, as in many other similar discussions, those who spoke on the occasion of such public meetings used specialized terms, words and categories that through the passage of time have become deprived of any content. Sometimes they even become formulations to camouflage erroneous assumptions. As an example, let us take a look at the expression "the role of the historical center." It presumes that there is such a unity as a "historical center." It presumes, socially speaking, a homogeneous whole that can benefit from certain measures. Indeed, we know that this is not so. On the contrary, is there not an ensemble of people constituting a number of social groups with specific, differentiated and often conflictual desires and problems? One would not have known that from the terms often used by the urban experts in Faenza (as elsewhere). Perhaps some of those who use the expression "the role of the historical center" do have this reality in mind. Obviously, however, that kind of phrase does not suggest they do nor does it indicate any capacity to communicate in particular detail such reality.

Perhaps for the experts in such professional rhetoric the quality of information exchanged in this way is sufficient because it conforms to the nonparticularized abstractness and superficiality of their working instruments. Surely, however, it does not communicate to nonspecialists whose real problems are thereby escaped or avoided.

Not by accident were the comments of the "average citizen" at the end of these public meetings generally of two types. Some underlined the point that they had heard the usual fancy words which, however, did not concern their everyday life problems. Others simply concluded that they had not understood much because "those things" were for the "experts," the "technicians," not for them.

Another traditionally conducted planning operation produced equally slight results. The municipal administration had to send the first planning document to the neighborhood councils, the labor unions, the business, cultural and other associations. Some meetings between experts and representatives of the most significant of such formal organizations were then organized. By way of conclusion, the administration received letters of comment on and reaction to the document from these organizations.

The first thing one must say is that there were few persons involved in this process, far fewer than its description would suggest. In fact, the "representatives" of the associations usually had done their job of representation by consulting with, at the maximum, their respective governing structures and not their rank-and-file memberships.

Despite the multiplicity of labels, citizen participation in such conventional forms was, in the end, a matter of the opinion of few persons. Such opinions, for the most part, were very restricted and specialized, since the various representatives limited themselves to examining matters presumably within the special competence of the respective organizations.

Only the document of the association of neighborhood councils urged the administration to find forms for enlarging the public knowledge of what was taking place and for allowing the people of Faenza to participate more in making the plan. Another association called "Friends of Art" was concerned with disseminating information and suggested to the administration that every family receive the initial planning document. Throughout this first phase no public attention was given to the social investigations proposed nor to the aspect of action-research involved. Since no major criticism had emerged, the city council also gave its approval and, thus, opened up the possibility of passing to the next phase where the social research was to be developed and begun. Very little if anything had been accomplished in this entire first phase because, as usual, the first planning document was controversial. It dealt with generalities, past trends, and at most began to imply something about future directions.

As the first step in the social-research phase a number of informal meetings were held with an ad hoc sample of people. These were generally long chats and conversations, eventually reaching around 160 persons, usually taking place in various public places such as the piazza, the open air market, and coffee bars. Such meetings were intended to generate a general impression of the life of the community, the daily problems and habits of the citizens. This knowledge was also useful in preparing a working tool for the next phase, the questionnaire or interview schedule, connected and conforming to the particular situations of Faenza citizens as much as possible.

Simultaneously, one of the most important and delicate tasks had begun: the recruitment and preparation of the group of local citizens who were to work in the core of the social action-research. This preparation is crucial in every social investigation, but it was even more so in the experience of Faenza for at least two reasons. One was that the group of so-called "interviewers" which was to be formed did not have the usual task of gathering a certain number of responses to a set of questions. Instead, it constituted the first nucleus of nonspecialized persons to be integrated into the group of professional planners by participating in the work of the latter. This was done as the first action of the prearranged strategy centering on demystifying, translating the jargon and opening planning to successively larger sets of ordinary citizens.

The other reason the interviewers' preparation was crucial was that their role extended beyond one-step data-gathering. After all of the information in the initial questionnaire was gathered, each of them had the task of speaking at least once more to the persons interviewed initially. The purpose of this was to get as many of the latter as possible to become the second, larger wave of participants and propounders of knowledge and interest in what was then unfolding. The persons making up the sample for the first series of contacts were understood to be more than people furnishing information on which we planners could base our work. They were seen as members of the community who, because of their relations with their families, relatives, friends and work associates were the crucial wave centers of propagation whose interests and energies we needed to activate or release.

It was in relation to these objectives that it becomes evident how vitally important the group of interviewers was. It also became clear that they could not be experienced, professional interviewers. It was important that they be members of the community, working there as naturally interested and participating persons.

The selection process led to the formation of a heterogeneous group of 15 persons, mostly young but including older persons. Some were students, others were recent graduates who had not yet found a job and some were housewives. All were selected partly because they themselves wanted to be active participants in an opened urban planning process.

The size of the sample for the initial interviews was set as a total of about 800 persons. Almost 600 of these were persons living in the urban center, that is, in the historical center and the surrounding suburban areas, and more than 200 lived in the rural agricultural countryside. The sample was drawn randomly from a complete list of inhabitants 18 years and over.

The local newspaper announced the start of the investigation and posters were also prepared. Then a letter from the mayor was sent to all those included in the sample. The positive reaction of the people was evident. Not only did a very small percentage refuse to respond (3 percent); almost everyone completed the interview, which was very long. Few gave casual or hurried responses. The great majority of those interviewed said they were available for further meetings and discussions. Given the importance placed on these conversational interviews, we proceeded quickly to tabulate, elaborate and interpret them. About a month after these interviews or conversations were completed, some of the important findings that had emerged were put at the disposition of the community.

Gradually, as the elaboration proceeded, these pictures and patterns provided the official urban planners an incomparably more complete understanding of the characteristics, aspirations and life of the members of the community than was possible with traditional data and procedures. But this is neither the principal point nor the most important aspect here.

What is most important is that, in the meantime, many members of the *community*, the citizens themselves, found themselves faced with various aspects of urban problems about which earlier they had been only vaguely and often imprecisely aware. Ordinarily, in fact, those who knew about or experienced the most severe problems regarded them as "personal facts" rather than "collective conditions." As citizens with such new insights were invited to enter even further into the processes of planning to try to cope with, diminish or solve some of those problems, their readiness to act increased dramatically.

The findings of the study were so important for the successive program of actions that, before speaking of those, it is important to have a general albeit summary picture of what was learned.

The Participatory Potential of the Community

One of the special features of the social research in Faenza, as mentioned, was that one could compare the picture of the historical center with that of the surrounding urban area and with the agricultural countryside.[13] Some of the differences resulting from this comparison were striking. As we will see, however, they prove to be more remarkable for contributing to our understanding of the modernizing processes underway in such countries as Italy than for contributing to a strategy for participation.

Briefly, we can say that the research has confirmed the condition of the historical center as a location with a much higher percentage of elderly people living there than in other areas. From the viewpoint of income as well as social isolation, many people were living there in extremely poor conditions. One fact clearly underlines

the difference in the phases of "development" between the two extreme areas, the historical center and the agricultural area: while
hardly any large, extended families exist any longer in the historical
center, they were found frequently in the agricultural area. Most of
the elderly in the historical center live alone or with only another
elderly person. In the agricultural area, on the other hand, they are
almost never alone; they live in traditional large extended families.

The differences in rural-urban living patterns also appear in
other ways. One difference that was surprising was that the fabric of
informal relations (relations with neighbors and visits with relatives
and friends) still seems to be substantial in the residentially much
more sparse agricultural area while it is much less extant among those
living in the historical center, who have absorbed much more of the
influence of, or been more subjected to, the so-called new models of
urban civilization.

Thus, the living patterns in the agricultural area evidence a
greater richness of social exchange and less individual solitude. By
way of contrast, those living in the agricultural area are present in the
formal associations of the community and in civic affairs to a lesser
degree than those living in the historical center and its surrounding
periphery.

To understand the significance of such patterns, we pass immediately to the condition of participation in the Faenza community.
Just as in Scorzé and Guastalla, the two Italian communities of the
international study, the number of persons present in organized
social activities and belonging to formal associations is low, but the
degree of participation in civic affairs in Faenza turned out to be
more influenced by class variables than it was in either Guastalla or
Scorzé. These variables included one's level of education, occupation
and income, as well as one's social background as indicated by the
level of education and occupation of the parents of those interviewed.

In other terms, the Faenza situation tends to be closer to the
American rather than to the Italian pattern described earlier, with a
substantial qualification. Although those who have higher occupational status, income, level of education and the like are present in
far greater proportions in civic affairs than lower-class citizens in
Faenza, such an upper or upper-middle class is numerically very small.
As a consequence, the middle and lower-middle classes are still the
most numerous in Faenza civic affairs, in contrast to upper-middle-
class dominance in American towns and cities.

In other dimensions the situation in Faenza is quite similar to those of Guastalla and Scorzè and also to the American community situation. The marginalization of the lower classes, for example, is remarkable; they are far removed from active engagement in civic affairs. One of the indications of class that appears more influential in this regard is one's level of education. This educational-achievement factor helps to explain the overall lesser presence in civic affairs of persons living in the agricultural area. In comparing the levels of education of people living in the three parts of Faenza we find a much higher proportion of persons in the countryside with the lowest levels of education. And that is the case also for young adults, although their level is higher than that of other age groups in the rural area. In other words, in the historical center the generally low level of education is not as low as it is in the rural areas.

This is important to keep in mind. It means that the cultural and everyday life differences between residential areas are, above all, due to sharp differences in such social stratification characteristics as the population's level of education. These differences go beyond those of life-style differences. For example, it is the difference in levels of education in the two areas that explains the existence of a much higher proportion of parents in the historical center than in the agricultural area who want their children to have a high level of education. In fact, in comparing the aspirations of people in Faenza by the educational level of their parents and by residential location, it is the educational level of parents and not rural-urban location that matters.

The large presence in the agricultural area of so many people with low educational levels thus determines many overall differences between areas. On the other hand, it is in regard to participation that the differences of mode of life between the urban and rural areas have an impact even if social and demographic factors are also important. In Faenza, the elderly and women, and especially housewives, were less apt to participate in civic affairs if they lived in the more rural rather than the more urban areas.

Women are generally less often present in activities and formal associations than are men. Only in associations and activities tied to the school and the church are women more often present than men. Those who appear to be most marginalized are the elderly and middle-aged housewives. This marginalization grows as one goes from the historical center through the suburbs to the agricultural area. A very low proportion of housewives, and especially of middle-aged housewives, is present in community activities and the activities they are present in are most often religious ones. Still, it is more difficult to speak of social isolation with reference to them than for comparable women of the historical center who do not even have extended families. The category of people who spend the most time by themselves, alone, on a typical Sunday and the greatest consumers of television both prove to be the middle-aged housewives of the historical center.

The situation of the elderly as a whole is much the same as that of the more isolated women. They tend to be nonparticipants in civic affairs, even more so in the countryside than elsewhere. The elderly in the agricultural area, however, are undoubtedly less isolated from other persons, since, as noted earlier, they at least enjoy the company of family.

Moving beyond these surely significant matters, let us return to the fact that participation in community activities (including those connected with local government and political matters) is not high and it is structured by such class characteristics as the level of education. Let us examine the state of affairs in Faenza in terms of such factors as cynicism and pessimism, the feeling of powerlessness, incompetence and the lack of self-confidence, which we proposed earlier as parameters of a model of nonparticipation. Then let us consider those elements that may indicate a potential readiness to participate and may be possible stimuli of such participation.

First of all, let us take a look at the data concerning the reaction that the people of our Faenza sample think their local government representatives would have if they came to them with problems. A large number, surprisingly large in comparison to the two Italian communities surveyed earlier, gave an optimistic response: that their representatives would listen and would do their best to resolve the problem. A further interesting fact is that there is only a very slight connection between this datum and class or demographic variables.

The elderly were not as optimistic as those of the other age groups; but they were not more pessimistic. Rather, they proved as usual to be more removed from civic life. They had a higher number of "I don't know" responses than did the other age groups. As for women, they were not more pessimistic than men. Nor does the variation of attitudes by residential area appear significant.[14]

With regard to one's feeling personally capable of influencing decisions regarding the solution of a problem, the situation was somewhat different. First of all, in the two Italian communities seen earlier the feeling of powerlessness appears to be a little more widespread than in Faenza, but in all three communities it is closely corre-

178   *Urban Self-Management: Planning for a New Society*

lated to indicators of social class; that is, with a lowering of social-class position comes a decreased feeling of being able to have an influence in community decisions. In comparing people of Faenza with those of the two other Italian communities in this regard and by educational level, the percentage of those who feel they have no influence is similar in the three communities for those of lower levels of education. But those of higher degrees of education feel more potent in Faenza than in Scorzè and Guastalla for whatever reason.

Most of the elderly in Faenza feel incapable of having decisional influence. It is well to recall here that the elderly not only make up the great majority of those with a low level of education but also the great majority of those with low income and low social status. All these are strongly intercorrelated dimensions and, thus, we see very negative effects on the feelings of the disadvantaged elderly about being able to influence community decisions.

The feeling of powerlessness, then, appears less striking in Faenza than in the other communities. On the other hand, on the measure of self-confidence or sense of adequacy, the three communities proved to be quite similar. Since the measure used in Faenza was a short version of the longer form used in the international study, we were able to make rigorous comparisons on the very same items. In Faenza, then, as in the other two Italian communities, people were comparatively very self-confident.

Such self-confidence in Faenza was connected to such class indicators as levels of education. It also varied with age. To what extent that variation by age was influenced by other, socially shaped characteristics of those of advanced age rather than by age per se, however, can be understood through an examination of young and middle-aged adults there. From that examination we find ourselves in total agreement with Simone de Beauvoir:

> But although old age, considered as a biological fate, is a reality that goes beyond history, it is nevertheless true that this fate is experienced in a way that varies according to the social context; and conversely, the meaning or the lack of meaning that old age takes on in any given society puts that whole society to the test, since it is this that reveals the meaning or the lack of meaning of the entirety of the life leading to that old age.[15]

What did we find? Generally, those who were the youngest adults, for example, below the age of 35, in Faenza and the other communities studied feel most adequate, most self-confident. We do find some young adults with little self-confidence. These, however, almost always have parents of low social class and have themselves low levels of education. We can infer, then, that it is those other factors that are more influential in producing low self-confidence in the elderly also rather than physiological aging. And of course they are a kind of social-fact indicator of a host of basic social and political and economic relations and experiences. Although we have too few upper-class or highly educated elderly persons in our samples to make an adequate assessment, we have sufficiently large numbers of middle-aged people to appreciate how very great are the differences in the self-confidence of the poorly and highly educated among them.

For women the variations in self-confidence are influenced by class variables and by their occupational roles. Working women do not differ greatly from working men in their sense of personal adequacy. We find middle-aged housewives and elderly women to have, together with elderly men, the most constricted feelings of adequacy, to be what we term the most cautious in their approach to the world. Only among young women, including housewives, does self-confidence appear not much lower than that of comparably young men. Incidentally, these findings are an interesting testament to what modern life must be like to affect so differentially the self-confidence of men and women—until the very last phase of life when men, too, experience the marginalization and disposal that we have prepared for our "nonproductive" human waste. We should also note that this sex differential is not due to educational differences because in Faenza men and women had comparable levels of schooling.

When we look at desires for more information as an indicator of potential readiness to participate we find that younger and more highly educated people do want more information, comparatively speaking. It is important, however, to keep in mind the fact that the actual number of those wanting more information in at least one major sector of community life is also high among the elderly, among women, and among persons with a low level of education (although comparatively lower than for others). Thus, this expression of interest shows that even these marginalized kinds of people are not as internally apathetic or insensible to community life as their nonparticipation or decreased self-confidence might suggest. They have been put on the margins of the modern urban society in Italy just as they have been in American communities but from these kinds of findings we know their spirits have not for the most part been broken.

In Faenza we had also gathered data on the readiness or nonreadiness of the persons interviewed to participate in meetings on various aspects of community life if they were assured their opinions would be taken into consideration. Furthermore, we had presented a number of proposals for action in the urban environment, proposals largely unlike the planning interventions traditionally proposed by urban planners.[16] For each of these proposals, we first asked the person interviewed whether or not he considered the proposal important. Then we asked him if he was ready to do something toward realization: give money, put pressure on the administration to implement the proposal or participate in even more direct actions.

The youngest adults showed the greatest degree of readiness to participate in meetings regarding civic affairs. In general, though, both women and many of the elderly said that they were ready to participate in at least one of the types of meetings proposed. And in every one of the three areas, the readiness to participate in meetings on problems of schools was high for all kinds of citizens.

As for the importance of the proposals to citizens, the findings once again attest to at least the absence of apathy or disinterestedness. Very few declared all the actions proposed to be unimportant, and many elderly people responded positively to more than one of them.

For our concerns with developing a strategy of stimulating participation, the findings about what people were prepared to do were relevant. We can set aside the proposal to give money because, as in the other two Italian communities, there are very few persons in Faenza who say they are ready to spend money—regardless of the cause. Instead, let us look at what we can consider readiness to participate in an indirect action (to put pressure on the administration in order to get it to carry out the proposals) and what amounts to an expression of readiness to participate in direct action (to participate in citizen action groups).

Both of these responses vary with social class. That is, as social status decreases, and especially if the level of education decreases, the readiness to participate in such actions decreases. The number of those who are ready to participate in the first type of action, influencing the administration, however, is very great and remains substantial also among the lowest classes. This is not true with regard to the second type, direct action. Few people are ready to participate in that manner and those with a low social position are almost totally unready. We have already seen that the elderly of Faenza, not so much because of their age but because of their social characteristics, constitute the prototype of these conditions. They are substantially present among those who are ready to participate in the first but not in the second type of action.

It is important to note that the social group that is clearly identifiable among those most ready for direct action are young adults of high educational levels, although not young adults in general. Those young people with a low level of education are not significantly present among those ready for direct action. The poorly educated, of course, constitute a much smaller proportion of young adults than of older people.

Putting together into a single index the responses to a large number of questions, we believe we have constructed a valid, reliable measure of the overall readiness of people to participate in community affairs in Faenza. We may look with profit at this measure in the context of the usual class and demographic variables as well as in the light of the factor that we earlier termed the sense of self-confidence (or vitality or personal adequacy).

The correlation between this measure of participatory readiness and class variables is once again notable, but the factor of self-confidence, the feeling of personal adequacy, has a remarkable function here. Let us take, for example, a significant sector of community life, that of local government and politics. Among those groups who are ready to participate in such affairs in large numbers are persons of young or middle age who have good levels of education with average incomes. In the light of a multivariate analysis of those in this situation—that is, in the light of the factor of readiness to participate and the factor of self-confidence as well as of the sociodemographic variables—a distinction emerges. There are many in this situation who are ready to participate but there are some who are not ready. To a very large extent this is attributable to the other factor, that of having a high or low degree of self-confidence.

This helps put in sharp relief the fact that even at very disadvantaged social levels we find in Faenza, as in the other Italian communities, a considerable number of persons ready to participate. We recall that the factor of "self-confidence" is influenced by such class variables as educational level. But in all three Italian communities a highly developed sense of self-confidence turned out to be quite widespread. Although it decreases with decreases in social positions, we still find substantial numbers of persons from the lower classes with a high degree of self-confidence.

What has been described here and, especially, the facts about

the readiness of people to participate permit us to draw two fundamental conclusions. First, the data that we have analyzed points to
the inadequacy of participatory efforts that refer only to a social
category, for example, the poor. Such categories do not admit real
persons, even conceptually or imaginatively. They are usually statistical, abstract categories, far from human and interpersonal meanings;
even level of education or income or occupation is a poor substitute
for such relevant personal characteristics as self-confidence, interests
in or readiness for participation.

These personal characteristics help to differentiate people and
their situations as well as aggregating others into categories. And
these personal characteristics are influenced by complex sociocultural
dynamics. In any event, they must be taken into account in building
an intelligent participatory strategy.

These results also allow us to take a further step forward. The
situation of Faenza and of the other two Italian communities shows
that the fundamental problem is not that of "creating" interest in
participation. (Similar conclusions hold for the North American communities as well.) For every sector of the community life which we
have examined, there exists a much greater degree of interest and
readiness with regard to participation than is now actualized or even
evident to the casual observer.

What shows through all the research is that people have a much
greater interest and are available to participate far more than they do
ordinarily. There is evidence of that in concrete experiences such as
the program recently begun in Italy in the educational sector: citizens
are now elected to participate in managing local school systems. We
can also learn another lesson here from that example, even if it is not
the moment to dwell on it. The interests of people can also rapidly
change forms as bureaucrats retake space in the decisional processes
that have just been opened. In such cases of bureaucratic reclosing,
the first to withdraw themselves, or to be excluded, are precisely
those belonging to the lowest and most marginalized classes, those
who more easily than others pass from direct participation to dele-
gating their rights to persons "more prepared."

The central element of a participatory strategy, then, is offering
a real potential for power relevant to everyday life in the community
to citizens ordinarily excluded from closed decision-making processes.
If those conditions are respected, there will be a substantial number
of citizens, even from the most disadvantaged categories, ready to
participate. The essential characteristic, then, becomes the desire

really to open the decisional processes to all, requiring as conditions of such participation only human experience and not specialized knowledge or expertise.

On the basis of the bits of knowledge that we now have of the community of Faenza, we can proceed to examine the experience of participation in making the plan for the historical center there.

Participatory Action in Faenza: Its Meaning and its Limits

After having seen some of the results of the beginning of participatory action, let us turn back for a moment to examine its meaning in regard to the persons involved in it. The initial intention was to make the interview a first act of participation. The fact that very few people refused to participate was due, in our judgment, to more than such favorable conditions as the publicity and the mayor's letter. It was also due, and much more so, to the fact that for each person the "interview" situation guaranteed two conditions: that of feeling adequate and that of feeling competent to participate.

The relation between the interviewer and the interviewee was one of equals. The person interviewed did not "have" to respond simply because the interviewer "had" to have the answer in order to complete the first part of his work. In fact, of the two it was the person interviewed who was the "competent" one. He was to furnish information about himself and his views, and it was he alone who was capable of expressing them. The fact of sympathetic fellow citizens' having come to the house of each person to be interviewed allowed even the most isolated and estranged to feel at ease and to feel that, at least this time, he was being heard. In other terms, our operations were similar to those of good survey researchers with a crucial difference: the interviewers did not think of themselves as such nor of the others as "respondents." They regarded themselves as students trying to listen respectfully to their teachers, citizens rarely or never before regarded as such.

But let us now take a look at the procedure followed from the disciplinary viewpoint, from the viewpoint of traditional urban planning practice. By so doing we can specify another order of reasons for the innovative potential of this approach. Suddenly we find a theoretical statement in urban planning, often repeated but always remaining a slogan, becoming a reality: the basic elements of the city seen as a system are persons, not physical spaces, blocks, buildings or services. And persons, all persons, can and should become active elements of the planning processes; they must become planners.

I would not like to be misunderstood here, however. I do not believe there is, nor is it my intention to espouse, a single methodological approach for planning that is better than others: a method that begins by carrying out a "good" social investigation instead of a traditional gathering of statistical data. In other words, I do not believe that recipes exist for how to make a "better" urban plan. The basic approach can be translated into a series of actions very different from these in Faenza. The important requirement is that whatever actions are undertaken anticipate and facilitate the greatest possible number of persons intervening in the processes of planning. And it is crucial to begin to understand that urban planning is not only locating services or deciding what the density of construction should be.

With that said, we return to the experience of Faenza at the time the data emerged from the social study. A synthesis of it was prepared and immediately made widely available. It was also sent directly to all those who had participated in the gathering of the data, most notably, those interviewed.

Meanwhile the interviewers were starting a second series of meetings with people to discuss their reactions to the findings and their ideas about what needed to be done. But these exchanges, instead of having the dimensions of traditional public meetings, had a small scale and occurred in all kinds of places in which citizens tended to gather. Such meetings had a refreshing character of vivacity, and of encouraging real contributions by those present.

An example will illustrate the very different meaning and the different consequences of this type of meeting compared with the traditional public one. People were generally surprised to find themselves faced with our findings and particularly with the image of the conditions of the elderly in Faenza. The vague or very particular, limited knowledge that each person had in such regards did not coincide with the dimensions evidently characterizing the actual conditions: deep economic poverty, often miserable and generally very bad living conditions, estrangement from social life, scarcity of social relations, and so on.

Another element of surprise, in this case positive, was that these problems were being approached and discussed as urban-planning problems. That is, they were being taken into account in the context of the preparation of an urban plan and not in the usual context of various kinds of welfare programs. Discussion of such matters began to make it possible for people generally to recognize the connections between the situation of particular citizens or even a part of the citizenry and the forms and fabric of urban life.[17]

The reaction of the elderly who were present in such encounters was in itself significant. Many clearly passed from a feeling of being spectators, who if interested were also prudent and quiet, to that of being protagonists. Aspects of urban life and problems appearing in the daily lives of other groups and persons in the environment of Faenza were also discussed as urban planning issues. This discussion also, but not only, involved the uses of physical space and the number of apartments, schools and other services that were felt to be needed.

Among those who participated in these meetings there was also a gradual emergence of persons ready to take part actively in further working groups. Such working groups would be formed area by area, and with the collaboration of the "experts," anticipating a further phase in the experiences of self-planning.

Some ideas became more precise proposals, not only concerning physical elements but also the management of such elements by the people. For example, the multifunctional use of certain public spaces for the purpose of allowing people of different ages and characteristics to meet was one such interesting kind of proposal. Another was the creation of vegetable and flower gardens to be used by sets of neighbors in the center of the city.

The Faenza experience took another route at precisely this crucial point. After a period of relatively slight intervention and of cooperation on the part of the administrators, things changed dramatically. Earlier they had collaborated in such things as publicizing documents and convoking meetings when requested. Now they took a new interest in events, evidencing their desire to take into their hands "the control" of the situation. This was made easier by the sociophysical division of work.

After the action-research concerning the community was carried out, a thorough investigation of the physical dimensions and conditions of all the buildings in the historical center was launched. The type of information that was to be collected concerned the traditional physical elements of urban planning: the number and size of apartments, building typologies, the artistic and historical characteristics of buildings, the location and classifying of economic activities, and so on.

(I must stress the point that the idea of carrying out separate investigations on the physical and the social aspects did not seem to me to be the best approach. In fact, it was absurd not to gather data all at once on the modes of life of individuals and families in the context of their physical space even if it was necessary to do so in a nontraditional form of large samples rather than surveying the total population. This separation into two domains of investigation at the beginning amounted to a necessary compromise within the group of appointed urban planners. Those in the group who felt that putting aside the traditional approach in order to follow a different working method did not guarantee results felt it was necessary to reduce the risks of being charged with failure by having in hand at least these traditional kinds of physical data and some other statistical census data.)

The procedure of two separate investigations did not constitute the best strategy on the basis of the proposed participatory goals. But it might have been an acceptable strategy if the data gathered in the two domains separately could have been reunited by bringing them back to the community and by considering them simultaneously in unitary discussions. Instead, when the results of the investigation of the physical aspect started coming in, the local authorities returned to the traditional procedure: a few large public meetings were convoked in order to "inform" the people.

In these meetings most of the local technicians, once again, returned to their traditional professional roles. These included maneuvering and manipulating by using such specialized terms as "building typology," which they could do far more effectively and in a more mystifying manner than could others.

The situation also evidenced similarity to American experiences; only the names of the protagonists were different. As I said earlier, the municipal administration was changed during the time we were doing our work and the new Communist-Socialist administration proved to be quite anxious about citizen participation that, in short, it neither directly promoted nor controlled. As a consequence, the political parties reacted. The strongest one, the Communist Party, as well as its much weaker partner, the Socialist Party, advocated new public meetings to present the physical survey findings quite apart from the results of the social survey. At these meetings, party officials and trade-unionists along with the local technicians assumed the roles of leadership. They did so by not leaving opportunities for common citizens to speak or, even worse, by treating as troublemakers those who sought to lead the discussion to the themes and problems discussed in the preceding meetings about the social and human problems in Faenza.

While in the process of writing what became the first, Italian edition of this book, I was still optimistic about what was unfolding in Faenza. Now, before drawing my conclusions nearly a year later, updating them according to what has occurred, it seems opportune to repeat my first comments here:

> The experience of Faenza has still not officially concluded in that the plan has still not been produced. It is not necessary, however, to wait for it to be finished in order to draw some conclusions. In fact, these can already be traced in the events which have taken place so far.
>
> It is my conviction that the openings which have been made within the planning processes will neither easily nor soon be closed again.
>
> There is a sufficient number of persons in the community who have begun to understand, and as a result to demystify, urban planning, its methods and its prescriptions. Through these processes they have also begun to raise fundamental questions about traditional decisional processes in the community.
>
> Although such a fact is relevant, it will not be sufficient to produce visible innovative results immediately, that is, in the context of this planning experience.
>
> What will be concluded in Faenza will be a plan which, for those who will not have the patience to read what I have written in these pages, will hardly show the traces of the goals set by at least some of those who participated in it.

The elements of ambiguity in the situation led me to judge the meaning and the lessons of the situation of Faenza in this way:

> The next and definitive draft of the plan for Faenza is only partially born from the direct contribution and work of the citizens. Initiatives that were particularly alive in some areas will eventually become the nontraditional forms of a "socio-physical" plan, at least for those areas.
>
> It is clear, though, that the format of the plan in its final official garb, and perhaps the organization of its management as well, will be predominantly marked by the usual routines.
>
> In the light of this nearly certain prospect, what has so far been achieved in Faenza may even take on the aspect of insignificant events.
>
> One must go back to the potentialities that were in arm's reach, but which for reasons now obvious were not realized, in order to have faith once again. One must intellectually confront other experiences in order to be in a position to establish the correct value of the results so far acquired

and the meaning of this experience. The first result is clear. No matter how the experience of the plan for the historical center is brought to a conclusion, when the four commissioned architect-urban planners leave Faenza, more than a technically beautiful plan will, in any case, remain behind them. It will be more than a plan whose authoritative and authorized interpreters were the few specialized officials of the municipality and/or the professional world of Faenza. It has appeared clear to at least some of the administrators that modern urban planning is not the best way to solve urban problems. There is not only a lack of adequate laws for increasing the "public" use of land. Modern urban planning does not and cannot cope with urban problems because it accepts the division of labor and acts according to its premises of expertise and of efficiency by also reinforcing the divisions of life.

Thanks to the events about which I have testified, there are now a number of citizens in the community who have discovered, and have continued to demystify, many of the secrets dealing with "how to do urban planning." They have understood that the organization of urban space cannot be dissociated from their social living patterns, from the problems of their everyday life. The fact that at least some administrators and citizens have acquired this awareness means that they will not be satisfied in the future with a traditional urban-planning instrument.

A second positive element is the fact that a part of the community of Faenza has, in the context of formal organizations such as party meetings, begun to discuss problems of isolation, alienation and the breaking of social relations. This is also taking place in their informal meetings, for example, in the coffee bars. Above all, they have begun to discuss these problems by identifying them concretely in the everyday life of their community. They are no longer discussed as abstract and distant problems from which only those in the large metropolitan areas suffer.

They have discussed, and perhaps have more concretely understood, that injustices are not only expressed through economic and social-status inequalities. They are also, and above all, expressed in the expropriation of the rights of all people to participate in the decisional processes of the community.

In other words, however little or much the experience of Faenza will remain engraved in the final planning product, it is important in having touched upon all three factors that prevent a greater number of those belonging especially to the traditionally subordinated and marginalized classes from participating in the social processes. These factors are: the feeling of pessimism and the lack of confidence in being able to influence decisions: the feeling of inadequacy; the feeling of incompetence.

Having remained at an initial stage, the actions carried out in Faenza do not allow one to hold that these impediments to participation were more than touched. The positive reactions, however, confirm those hypotheses set forth in the preceding part, which concerned the strategy for getting those who belong to the most marginalized social classes and categories to participate actively.

To obtain more participation, especially among such citizens, one should not opt for information that generates interest, which in turn constitutes a push toward participation. He or she should opt for the direct actualization of participation because participation itself constitutes a blow to the three abovementioned barriers to participation. Pointing out the positive elements may give to those who have gone through the experience a small degree of optimism, which is a necessary support if the number of those who feel pessimistic and impotent is not to increase. It is even more important, however, to see the reasons that have prevented the experience of Faenza from developing further, from becoming even more meaningful.

Precisely because of the importance that this census of the negative or at least diminishing elements has, I will proceed to a detailed and systematic exposition. . . .

I will stop here. It is not worthwhile to report the detailed analysis done then. I think that the preceding parts have given enough of an idea of what the situation was like at that time. Since then, several events have occurred that throw new light on all that I have written, not only in the sections on Faenza but also on my fundamental thesis: that of the importance of urban self-management, of despecializing the disciplines that deal with urban organization and the problems of the city. I mentioned before the reaction of the parties that now control the administration in Faenza, the Communist Party and the Socialist Party. The administration tried to take in hand the public meetings and to bring them back into the context and form of the usual meetings convoked to "inform" the people.

My evaluation at that time, which clearly appears in the preceding remarks, was that the administrators, or at least some of them, had understood the importance of what was begun. Although not having the "political will" to modify the products that urban planning must legally produce, they intended to take account of the interest aroused in the community and the readiness of the people to participate. Or so I thought then.

It seemed to me patently obvious that the two parties of the left, even assuming their desire for more direct control, could not oppose efforts directed to finding ways of actualizing interests in participation in decision-making on such community problems as urban planning. In a word, these parties of the people, of the most suppressed portions of the public, would favor initiatives seeking to arouse the "political conscience" of the people. These assumptions proved to be wrong.

In fact, the initial reaction of the new administration became, more and more, a real counterattack. Initially it took the form of putting heavy pressure on the team of urban-planning professionals commissioned to do the historical-center plan, who were called to account and had to defend themselves for not having done the traditional planning work with sufficient speed. They were accused in an increasingly explicit way of not knowing how to do their work, of being incompetent. Not without significance is the fact that just then other urban planners were commissioned to prepare a new master plan for the entire city and county. Persons closely connected to the two parties now in power were chosen as the professional urban planners, a normal procedure in Italian municipal government practice. The planners who were so commissioned were known as efficient and able interpreters of urban-planning procedures and legislation, valid "experts" of the administration.

It was to them that the task was entrusted of "coordinating" the plan of the historical center within the more general work of revising the master plan. This they did by filtering and absorbing the center's problems, the concrete problems of actual people, through the usual technique of melding them into abstract and statistical calculations involving the entire community environment.

An example is the situation we found of solitude and isolation, the living conditions of most of the elderly and of other poor people in the historical center. These become facts no longer worthy of consideration if read by means of the usual statistical alchemy of age pyramids constructed by the planners for the inhabitants of the entire municipality.

At the same time as this attempt at "technical control," another process began, an attack against the proposals finally produced by the urban planners for the zoning of the historical center. Those proposals first of all took account of needs advanced by people and indicated spaces and buildings in the historical center for community use, proposing that they be self-managed by citizens. Opposed to this was the so-called "realist" line, preoccupied with the reactions of property owners and, as always, reckoning with the proposals only in economic terms.

The particular situation of Faenza was such that the property owners to be defended were distinguished religious institutions. This did not mean that only the opposition parties and particularly the party traditionally associated with the Church in Italy, the Christian Democratic Party, mobilized the defense. Instead, the "realistic" defense of the property owners and the attack against the "utopian" planners were carried out by the parties controlling the administration, above all, the Communist Party. As in Italy generally and, indeed, in much of Western Europe, so-called Eurocommunism needed an image of being a supporter rather than a destroyer of private property.

Thus, the situation is now quite different from that of about a year ago. There is now a distinct possibility that everything may be canceled. There is a possibility that the aims of "revitalization" of the historical center designated as the starting point by a theoretically moderate municipal administration may be negated. There is more than a possibility that these aims may be relegated by an administration theoretically of the vanguard to a category of populist-cultural dreams.

It might also happen that the formal commitment to a plan for the historical center will be maintained. This time, though, it will be to produce a traditional plan in a climate of counterreform. It will allow no one except property owners—that is, not the bulk of the people who are tenants—the right to use land or space as they see fit, subject to general rules of respect for others. For the twin objectives of improved urban quality and equality it is vital to open the alternative planning process to citizens regardless of their property status but in Faenza that now seems impossible.

It would not be important to have dwelt in detail on the events of Faenza here if they were not able to furnish us with very general lessons that go beyond the Italian situation, lessons concerning possibilities and strategies of action in other urban milieus of modern society.

In the light of the growing strength of the parties that "represent" the disadvantaged classes and given the constant rate of increase in support for the Communist Party, many political analysts now anticipate real and profound transformations in the Italian situation. Since the elections of 1975, most major and many small Italian cities are now governed by coalitions made up of the two parties of the left, the large Communist Party and the much smaller Socialist Party. Many of the regions are also now governed by such a coalition. Despite such facts, the behavior of these parties in the case of Faenza is emblematic of the fact that no major innovative aspect is present in the parties of the left in Italy.

The experience of Faenza confirms the fact that the structure of the Communist Party (as of the Socialist Party, even if less formally) is exclusive, hierarchical and centralized and in the same spirit as all the major modern institutions. In their operations in Faenza, as in Italy generally, the parties of the left have tended to close decisional processes just as much as have the parties of the right. Under their direction, decisional processes, in fact, have been increasingly thought of as being within the specialized competence of the party. In the meantime, however, within the parties it is not the majority of the members who make the decisions. It is those who have responsibility, the leadership, the "representatives'" of the rank and file, who decide. The process of opening decisional processes, of "democratization," is seen more than ever as a simple matter of increasing the number of party supporters, according to party logic.

In fact, more and more aspects of social life are becoming matters for "political decisions." This does not mean, though, that it is understood that everything in the social domain has a "political" meaning. Nor does this mean that there is an ever-growing conviction that nothing can be "technical," "scientific," "objective" or "neutral." The parties of the left have as much if not more respect for technicians as do the conservative parties.

What is happening in the present phase of societal evolution is that a process of reconcentration, or if one wishes, rationalization, of decisional power is in course. Previously, this decisional power was gradually fragmented in a myriad of institutions for culture, education, sport, health, and so forth, which comprised the community's associational life. Now this reconcentration and reunification of decisional power is taking place in Italy around a party of the left, the Communist Party. But this is far from guaranteeing renewal, apart from the circulation in the identity and political orientations of those holding decisional power. It is not an opening of the decisional processes.

The leadership of the party that by definition "represents" the dominated classes is now spreading and stabilizing and in the process there is a propagation of the belief that the masses automatically, without needing to make personal contributions, are thereby present in decision-making, in positions of sharing in power. This is the crucial point of the illusion of political transformation that countries such as Italy are now living through. The truth underlying this illusion is demonstrated by the reaction—which otherwise would appear out of all proportion—that exploded in Faenza when an administration of the left was carried unaware up to the point of what could have been a really major happening of urban self-planning.

Because the mistakes are so obvious, not many words are necessary about the principal errors of those, including and above all, the writer, who participated in that process with consciousness of their innovative objective although not of the detailed ways and means of achieving it.

There are useful lessons to be learned from the entire occurrence at Faenza, especially for those in situations with different characteristics from the Italian one. The Faenza experience underlines the need for more appropriately defining the objectives and instruments of action. Above all, however, it helps in clarifying the social forces and groups whose contribution is so essential that it cannot be renounced. And Faenza makes clear the imperative of having at the outset a set of professionals willing and able to engage in professional demystification and opening the process to nonprofessionals. These professionals must also have a strategy to cope with the political party and other institutional forces that, despite rhetoric, cannot but oppose as personally threatening and ideologically and philosophically wrong any real and substantial institutional openings and transformations from "expert" to self-managing processes of planning or decision-making.

TOWARD A THEORY OF URBAN SELF-MANAGEMENT AS A PROJECT FOR TRANSFORMING SOCIETY

Society and the City

In the light of what has been said, it now seems possible to return to the principal arguments around which this book has been shaped. By doing so, we may connect those to the unfolding of the experience at Faenza, especially its last phases, as well as add some reflections that the experience suggests.

Let us begin from the field in which the analysis was located and in which I propose to focus actions: the urban environment. Further, let us begin from the relations between the city and society about which we spoke in Chapter Two.

We have seen that society must be read as a system in which there are organizational forms called institutions (political, economic, social, cultural), which human beings have shaped and which they maintain in operation. To these institutions are ordinarily attributed superhuman qualities.

The underlying reality of the presumed superhuman quality is, on the contrary, precisely the fact that all institutions are no more and no less than groups of human beings. It is a reality characterized fundamentally by the presence of particular people belonging to certain classes or social groups and living, performing or experiencing in symbolically distinctive ways. In other words, institutions are "exclusive"; those who are "inside" are differentiated from those who do not have such membership rights. The underlying reality is also characterized by a hierarchical organization that reflects and effects differential situations of power even among those who are members of the same institution.

Every institution expresses a political line even if it is not called a political line. That line is elaborated by those men who control and guide the institution's power structure and is given substance in a network of decisions programmed and carried out on a scale that in some cases includes the entire society. In other instances, that network of decisions involves a set of cities or a region; in others certain specific cities.

As a result of this, the city cannot be considered a microcosm or a subsystem representative of the total social system. In fact, there is not a single city that is a microcosm containing in all of its aspects the type of power structure shaped by the various institutional presences characterizing society.

There is also a further element that makes the city and society nonhomologous. The city constitutes a "spatial node" of society; it is the place in which institutions are given existential reality by acting human beings. Thus, they must respect an essential human condition, that of developing in physical space. Otherwise institutions would be abstract, nonexistent units.

It follows that the city is more than a simple location of institutions in space. It is the place where people have potentially the power to demystify the presumed superhuman quality of institutions and reshape them. This potential power has until now gone nearly unrecognized by most people. Yet, it is this element, although different from place to place, which is common to the form of human aggregation that we call the city. It is less concentrated and is in more diluted forms in the still rural but increasingly urbanized places.

In the real world, in terms of actual rather than potential power, the opposite situation exists: it is in the urban environment, not the rural, where most human beings, rather than expressing their ability to be influential, are increasingly alienated from their decisional capacities. That is, it is in the urban environment that the decisional power of most people is progressively and increasingly expropriated.

But let us look at the processes through which this condition characterizing the modern world has taken on importance and consistency. We find that one of the principal reasons for this development is that the dominant class that guided the industrial revolution has succeeded in imposing on the other classes the division between work and life or between the life of work and social life. (This is not to say that the dominated did not welcome the domination.)

Limited by an ideology accepting the divisions, the class struggle, even where it has been the most bitter, has not focused on them. It has aimed instead at obtaining the progressive recognition of the rights of workers within the so-called productive system, rights such as better working conditions and higher pay. In the urban environment demands have also focused on obtaining better and more widespread services.

Neither one nor the other of these conflictual processes of demand-making touched the bases of the power structure, of the traditional decisional structure. It was precisely in the urban environment, in fact, that the dominant class more easily found a way to recover the possibilities of dominating, some of which were lost in this century by victories of people struggling to organize unions and the like. This reconcentration of power, of domination, occurred despite the development of so-called political democracy with such features as the right to vote extended to everyone.

In the second chapter I spoke of the greater mobility, the reduced restrictedness to precise locations, that major power groups manifest, especially in the present epoch. This, nevertheless, does not mean that it is now unimportant or unnecessary to control the areas in which these institutions, and the people who comprise their reality, must realize themselves in space. This greater mobility or presumed flexibility of location is only partial.

By way of explanation: an economic power group can usually find another city and always another country that may guarantee more secure, more profitable and simpler conditions for its operations. That may happen if in a city, or even in a country, sufficient obstacles are put in the way of the group's initiatives or if its prerogatives are called into question. Under some extreme conditions, of course, coup d'etat or counterrevolution may be the reaction of economic power groups if the costs of moving are regarded as too high (Allende's Chile or Cuba as Castro was moving towards power).

And it is true that no matter where it is convenient to organize production, the location of consumption, of markets, must be taken into account (at least thus far and it is difficult to imagine otherwise). This consideration has demanded and demands constant control over the cities and urban places that constitute the major markets.

Such control conflicts at least theoretically with the formal development of political democracy noted above. The control is exercised by fragmenting the social sphere into pieces and by isolating human beings. In this way a person can more easily be conditioned, molded or set apart, precisely because he can be influenced individually. We have seen that entire categories (women, the elderly) are nearly totally marginalized. We have also found how scarce participation in community, civic and social life generally is. It is well known that the fabric of relations between friends, neighbors and even relatives is weakening everywhere in modern urban societies.

But there is still another fact to underline. Our data analysis was addressed to the small community. This shows that, contrary to what is often asserted, alienation and isolation are not only phenomena peculiar to the large metropolis. The size of the city can have effects on or reflect the more or less acute forms that these society-wide problems take; it cannot affect their substance.

Given the scope of marginalization in urban society, this would seem to cast doubt on the validity of a strategy that begins from the cities in order to transform society. Instead, by a double evaluation, both of the negative present situation and of its positive potentialities, this conviction that this strategy is sound turns out to be reinforced. Furthermore, the goal and the instruments of action to reach the goal also appear to be confirmed.

Let us try to sketch the heart of this apparent contradiction in which the potential situation is the obverse of the actual situation.

1. At least in the capitalist world, the city is the place in which the subordinate classes undergo the most expropriation. The exploitation of these classes in the so-called productive sector has been conceptualized for some time. What remains incomplete and vague is the recognition of the expropriation suffered by these classes through the degradation of the human quality of urban life, through the conditioning that these classes undergo in regard to the texture of social relations and of their own human vitality.

I do not want to contrast a romantic vision of rural poverty with that of the urban scene. I do believe, however, that one can speak without fear of contradiction about nonurban man as a man who is less conditioned and distorted by partializing divisions of life and labor. But nonurban man is gradually disappearing (another reason for accepting as valid a strategy of transformation that begins from the urban milieu).

2. The city is also the place where the greatest change is possible. The present situation and the position of the dominant classes now in control are unassailable due to the lack of aggressiveness of the dominated classes. The latter have not pushed for such themes as the

right to decide——for those very things that we have called the human qualities of urban life.

At the same time, though, the traditional power groups in the urban environment are more vulnerable than those in the area of industrial production or production in general. The reason for this is that the power groups in the urban milieu express divergent, and at times even conflicting, interests. The proliferation of institutions has created a further fragmentation in social classes by introducing a number of privileges that have created a much more subdivided and less rigid stratification of classes. This has made it, and continues to make it, increasingly more difficult for those who traditionally have decision-making power to operate on the basis of a stable, fundamentally simple class solidarity, on the basis of informal gentlemen's agreements.

The Significance of a Strategy of Urban Self-Management

To single out the urban environment as the focal point for a strategy of transforming society, we draw attention to the conditions of alienation and marginalization, the expropriation of decisional powers, and decisional deprivation as our starting point. It is these conditions that characterize urban man and that are at the base of the problems of modern society.

In turn, the nature and objectives of the process of transformation that can be activated in the urban environment require one to consider the "forces" and consequent forms of action at one's disposal. In certain urban realities, as in that of Italy, the presence of organized political forces, of strong parties of the left supposedly representing the interests of the have-nots, seems to constitute a huge advantage, at least at first glance. Actually, however, even the strong political presence of left parties on the social scene constitutes more of an obstacle than an advantage for transformations in which power does more than simply change hands. The experience of Faenza has brought this into relief and, as I have already emphasized, so has the very structure of the "institution" called a political party, even if on the left.

In comparison, the North American reality seems to present greater potentialities for transformation since it is faced with the deepest and most widespread urban crisis (but not only for that reason). And that crisis is especially acute in the United States. Such potentialities have been touched upon in a minimal and partial way to this point.

The most significant proposals that have matured during the most recent historical period can be categorized and summarized in the following way: they urge either more services for the disadvantaged, more direct protests by the disadvantaged or more political power for cities and thus more services for the disadvantaged. We will consider them in order.

The first group is one of liberal-reformist proposals of the left, such as the Cleveland example mentioned in Chapter Three, in which urban planners put themselves on the side of the poor, the segregated, the disadvantaged, in order to secure for them more public housing and better services in the plans. It is worth repeating that such a position cannot successfully meet the problems of the city and society. It could not be successful even if the problems derived only from the economic poverty of even a large part of the urban population.

In fact, this position does not touch the crux of the problems of modern urban society. Once more, it evidences an approach based on the ideology that has led to these problems, an approach shaped on a model of man as quantifiable and divisible into a series of needs, which needs can be satisfied by particular need-specific, specialized services that some provide *for* others.

It is true that in order to survive physiologically, man "needs" a minimum amount of certain goods such as food, clothing and shelter. It is just as true, however, that in order to live humanly, he needs to make or do or have certain experiences beyond the mere satisfaction of animal needs that pertain only to the level of his physical survival. Eating, for people, is a human experience. Only a process of brutalization can reduce it and other needs to an everyday economic operation aimed at physical survival.

In elaborating the notion of services as the satisfaction of needs, bourgeois society has annulled the dimension of "human experience." Yet, it is precisely this that distinguishes the satisfaction of the needs of man from that of animals. Furthermore, in being directed toward "satisfying" needs, rather than toward creating human experiences, services contribute to man's diminution and alienation. This is true of man both as a producer and as a consumer of services.

One therefore does not set up an alternative to the actual situation by fighting for more services, which actually is a way to guarantee survival of the basic contemporary reality by touching only economic aspects of the social injustices present in contemporary urban society, and even those indirectly.

The second group of proposals, although more dramatic, is equally inadequate: it urges people to direct actions of protest and of civil disobedience when faced with urban crises. The so-called self-reduction in the cost of services is a practice that has been used for some time now in European countries (especially in Italy) by groups referred to as extraparliamentarian.[1]

Those who have just lost their jobs as well as those who have been unemployed for some time and other have-nots need somehow to make their voices heard in order to obtain a job like the one they lost or have welfare services restored to earlier, precrisis levels, but our judgment must be similar to that offered in the case of liberal-reformist proposals because it is certainly not a major advance to return to earlier alienating conditions.

And if such proposals are intended to speed up the final crisis of capitalism, one can only consider them simplistic and self-destructive.

I would like to emphasize that these forms of action are not inadequate because they refer directly to certain categories of persons; they are inadequate inasmuch as they fragment the various claims and renounce attempts at direct action to open the hierarchical structure and processes of the present decision-making system.

It seems important, however, to dwell more at length on the third group of proposals, several of which seem on the surface to be similar to those proposed so far in this book. The first of these comes from those who hold that the city should orient its battle to obtain more decision-making power vis-à-vis higher levels of government.[2] The others pertain to the criticisms of the various programs of the sixties in the so-called "war on poverty," and especially the Community Action Programs. Although there are some verbal differences, the latter critics agree in stressing how such programs were shaped by the traditional approach of merely giving more services to the poor, rather than by arranging for their active participation and the opportunity to self-manage such services.[3]

What makes these positions different from what I have been proposing? Several fundamental elements. First of all, let us consider the fact of posing as an objective the state's "transfer" of more power to or sharing more power with the city. This does not amount to a truly innovative proposal. To the extent that it is only a displacement of power from a higher institutional scale to a lower one, it cannot transform society. Recent instances of decentralization in Italy from the central to regional governments demonstrate this.[4]

Furthermore, some such proposals emphasize the fact that local units of government need sources of economic financing not only deriving from taxation but also from an entrepreneurial role. Surely, though, even an innovative local public enterprise is not seen as a means for stimulating citizen participation in the management of such enterprises. Attention is also addressed to resolving the financial crisis of the cities when some decentralization or devolution or tax-sharing is proposed. Thus, the emphasis is on putting the cities in a position where they can perform their duties in a better way, including assistance to the economic poor. The decision-making poverty of people is not often if ever an element in such concerns.

As for the other position in this group, I fully share the entire criticism of the "war on poverty" programs, but I am not convinced that a sufficiently valid alternative emerges from the critics on the left, not to speak of those on the right. Let us take a look at the reasons why.

Indeed, the poor must be allowed to manage the initiatives and programs intended for them. Only in one instance involving Community Action Programs did the poor elect their own representatives who were responsible for drawing up initiatives. In all other instances it was a question of representatives of the poor being nominated by politicians. But the nomination of representatives is itself insufficient. One should, instead, think in terms of organizational forms that involve actively those belonging to such disadvantaged social classes in a much more direct way and in a way that begins to open not only governmental but also other institutions.

Such an approach naming representatives, then, is entirely insufficient if not generally mistaken. The effort that must be made is that of helping the poor and other marginalized and subordinated people to obtain the opportunity to manage not only initiatives especially meant for them but also to participate in the decisional processes that concern the entire community. And those opportunities necessarily must involve institutional openings and transformations.

Of course, it will be immediately pointed out that the poor have little or no institutional competence or credentials; they have hardly any education and a good number of them have difficulty in expressing themselves. (I do not mean that this will be mentioned by those whom I have indicated as belonging to the preceding position but by those who find those proposals too avant-garde or merely bad.) Such objections are pointless. They are pointless insofar as they concern the organization of urban life, in which decisions have assumed, and quite unnecessarily, a specialized character that even the highly educated not versed in the particular specialty cannot understand. It is also clear that the innovative significance of opening urban planning processes is lost if that means to involve people rather than the so-called experts only in decisions on land use.

It is also insufficient to propose, as do some of the so-called "public choice" advocates, that since public goods and services are diverse and diversifiable, people should be able to choose among them as people assertedly may in private markets.[5] People need to enter more into the ranks of producers of such goods and services and into the ranks of planners and managers as well rather than merely become better-informed or even more influential consumers of goods and services provided and distributed by others.

To speak of getting the poor and others to participate in urban management, without repeating the preceding lengthy discussion, means to begin to question the divisions of urban life according to so-called functional institutional definitions. It means, among other things, to go well beyond suggestions and programs of decentralization, of local control, to efforts at redefining and opening schools as well as factories.

Among proposals that seem innovative but in our terms are not are those that involve poor people helping other poor people or in having skilled elderly people, for example, repairing things for the unskilled elderly. They do not address the central problems of opening institutions. I think that the difference between such proposals and those carried forward through this book will prove clear.

My proposal for urban self-management, however, raises very large questions at the so-called political level. On the one hand, probably no one has seriously proposed until now that the management of the city should be entrusted directly to the poor—together with other citizens. On the other hand, the hope that such an approach would succeed is clearly questionable given the likelihood of its encountering much greater hostile reactions than did other, basically more harmless proposals.

The overall failure in Faenza, a small Italian city where the interests at stake were really rather limited, should be sufficient proof of this fact. My reading of other realities, however, especially the American reality, leads me to the opposite conclusion.

Maieutic Urban Planning and its Protagonists

For the moment I will go beyond the problem of how to increase people's interest and, above all, how to secure the active participation of people and especially of lower-class and marginalized people. This is not a secondary problem, but once conditions exist that guarantee the possibility of *effective* citizen participation, it becomes a simpler one. We have already seen that if conditions that, for example, enhance people's feelings of adequacy and of competence regarding their ability to participate are respected, they will at least be interested in participation. This includes the most marginalized strata, even though proportionately these strata are less ready to participate actively than others.

Here, then, we enter into the merit of other crucial questions, beginning with the one concerning people belonging to other classes: why should they accept initiatives that include the lower classes in decisional processes? Usually a system of representation puts upper-class interests first. Why risk that? Why should they accept the proposal of acting together with everyone? There is also the question of why local administrators should consent to sharing their power with all of the people by proposing such initiatives as citizen participation in urban planning.

Finally, why should urban planners accept the idea that others, average or even above-average citizens but without expertise, as well as the poor and ignorant, are capable of making decisions in sectors where planners have special competence? Why should they make themselves available to ordinary people and commit themselves beforehand to whatever emerges?

The answers are not simple. Let us leave aside ideological reasons and humanitarian motives, presuming that if there are persons so motivated, certainly they cannot but help; instead, let us examine these matters starting with the problem of the readiness of citizens of other classes to concur with such proposals. I do not, of course, expect the privileged citizens of the highest classes to agree. But within various classes in the middle there are people who, despite having lots of money or little money, may or may not be in managerial circles. Just below the elites are masses, circles and strata of people who have little or no decisional power; and in modern society, even those with some power often have or exercise it only in limited domains while in others they are quite impotent.

Other conditions and types of privileges differentiate middle, high or low levels within these large categories. While there is mobility, the passage from being powerless to being powerful is a long and usually difficult one, if and when it is possible. This is another way of underlining the "proletarianization" that the middle and lower-middle classes in America have experienced that is concealed by the relatively widespread economic well-being there compared with other countries.

Our contention is that many people in these middle- and lower-middle-class circumstances can be made to understand that in their present state they are substantially powerless. They can be made to understand that the only way to get power is by rejecting the present illusion of prestige. This illusion derives especially from looking downward and feeling—by comparison—above the disinherited masses who, at an economic level, make up the have-nots. But such illusory prestige is not an indicator of power in modern America.

People of the middle classes need to renounce pseudopower and seek real power through strategies of de-alienation and by reappropriating their decision-making capabilities. They must understand that only in this way can they answer the malaise of the young, their own neuroses, and many of the other problems weighing down the middle classes especially, a weight due precisely to their decisional impotence in so many domains.

The problem with respect to the representatives of local government and the local administrators is not different, even if its dialectical character is more evident. They are led to believe that it is the present system that maximizes their power. It is in this situation that they constitute the legal spokesmen of the community and have the power to make decisions on its behalf.

To the contrary. It is precisely by enlarging the decision-making context that the role of these representatives may become more of a commitment and more meaningful. In fact, if one looks closely, what does their power amount to today? We will not even consider such personal privileges as having an official chauffeured automobile at one's disposal. Instead, let us take a look at the sum of limitations and conditioning factors that they undergo. When we do that, it is clear that the expropriation of their power is in fact equal to that suffered by the entire citizenry. They have become so constricted and convoluted in their professional and political performances by the advanced divisions of institutions that it is little wonder that we hear so much about a crisis in governing in the Western democracies.

For it is in those Western societies that modern urban society is most advanced and decisional powers and processes are so fragmented.

There cannot be a meaningful decentralization that results in more power to local government if people do not first reappropriate their decision-making "power." When, and to the extent that this occurs, in fact, the people themselves can concentrate power at the community level. They need not expect nor wait for altruistic donations of power from on high. Through coordinated action, people can take democratic control of institutions within their own territory, that is, in their own urban environment. By doing so, they will be carrying out a democratic transformation of society. Other forms of strengthening institutional control can be obtained but their effectiveness lies in their dictatorial rather than democratic character. It is certainly not such centralizing events that strengthen the meaning and role of the representatives of local power, as modern dictatorships demonstrate.

Still another question can be raised, however. How can one talk about transforming society if only the local government is ready to launch initiatives of citizen participation? Can anything result from a local government's being willing to share power with the people at this institutional level where there is already so little power? Starting a process of urban self-management means that all of the institutions in an urban environment and not only local government must be affected. If only one city took this route, it would indeed be easy for the traditional forms of power to deal with such an obstacle. It would, however, be much more difficult, given the multiplicity of fronts that would open, if an increasingly greater number of cities would do so and on a broad institutional front.

The representatives of local power therefore have two basic alternatives. They can continue to carry on their shoulders the huge weight of increasingly deteriorating situations: in this case they will actually become more and more insignificant while, perhaps, maintaining the appearance of being men of power albeit increasingly little. Or they can renounce this pseudopower to become coordinators and members of communities that are vital, dynamic and influential because they are composed of persons having such characteristics and relations. The alternatives presented to urban planners, and to all those functioning as "specialists" in the urban milieu, are identical to these two.

The current crisis of particular cities, the general theme of cities in decline and the scarcity of resources for substantial urban improvements are thoroughly documented. In that context it seems that the increasing importance of the role of urban planners and of all urban specialists is being recognized. In this case too, however, there is more appearance than substance. In fact, it is more imperative than ever for those now in power to regain control over the urban situation, to impose technocratically more efficient and ever less human forms of organization on it. In such circumstances the role of urban specialists, and especially that of planners, can become even more important. This can only take place, though, if and when specialists are ready to act as interpreters and agents of this need for control, for conditioning urban life. In other words, they must be ready to carry out the will of the dominant people in an even more efficient manner. They can become more important as planners, as professionals, as quasi-machines, and less important as people.

Alternatively, these specialists might help to broaden decision-making processes in urban organization, they might agree that the goals of social effectiveness should be defined through participatory processes, and they might begin to despecialize urban planning.[6] They might take the lead in suggesting how use values might be attached once again to physical spaces to supplement or substitute for the profit-oriented exchange values that now predominate.

The fact that the urban planner and other urban specialists should perform stimulating, supportive and animating roles right from the beginning should not be concealed. But the most important aspects of such leadership roles and the most innovative aspect proposed is that of demystification, of professional self-demystification. And there is nothing wrong with leadership if its goal is its own destruction by or submersion in a broader self-managing leadership. Nor is there anything wrong with a goal that can never be finally reached but only approached infinitely closer through imaginable time.

The role of which I speak is a maieutic role, a term derived from ancient Greek philosophy. According to *Webster's Third International Dictionary*, "maieutic" is an adjective that is defined as "of or relating to the dialectic method practiced by Socrates in order to elicit and clarify the ideas of others." We might only add, "and his own ideas."

Through this different type of planning, *maieutic planning*, people may become aware of their capacity to make decisions, of their ability to participate in community decisions. It is through maieutic planning methods that people of all classes, groups and strata can more easily recognize their own competence. The planner's role becomes centrally one of helping people recognize how planning concerns natural facts and actions performed daily, simply by living their everyday life in urban space—as well as the more unusual decisions that need to be taken periodically.

The urban planner now faces the following alternatives. He can continue to use obsolete instruments, methods and models that so far have been ineffective and that tend to become ever more technocratic, ever more removed from the people for whom it is said that they function. Or he can open the processes of decision to the people themselves. It is precisely the participatory aspect that marks as innovative, changes in the contents and methods of the discipline of urban planning. But it is, it must be, participation that is really participation and not programmed pseudoparticipation constricted to limited actions by people who are essentially outsiders.

Paradoxically, perhaps, we said that it is in the United States more than in such countries as Italy that we think the introduction of maieutic planning may occur first. We say paradoxically because it is precisely there, in the United States, that the professionalization of urban planning has proceeded faster and further than elsewhere and in the context of the most advanced divisions of labor and life of any mature industrial societies. Moreover, it is in the United States that the so-called urban crisis is deeper and certainly more evident than almost anywhere else. In such circumstances is it not to be expected that even more professionalization and more technocracy is to be expected rather than an innovative opening up?

Because the situation there is so much worse than elsewhere, or so much more difficult, and past professional approaches and efforts have proved to be as fruitless as the other kinds of reformist approaches mentioned above, planning has a chance of becoming maieutic and being opened. In the light of professional planning's bankruptcy there, a new effort involving real citizen participation might be initiated. The desperation and frustration with prior approaches that is felt by all the major actors, as well as by the passive audience, may make the concept and operations of maieutic planning appealing. Compared with Italy, the United States does not have so heavy a hand in hierarchically organized, centrally oriented political parties, to which parties the Italian professional urban planners look increasingly for their patronage. Political party weakness, inefficient and localistic organization and public mistrust combine to create a situation of opportunity in the United States.

But, the reader may ask, is it not in Italy that the political parties and movements of the (more radical) left are much stronger? Is it not in Italy that some of the direct actions have occurred in cities that make any suggestion that the United States is a more likely first arena merely theoretical and actually silly? Is it not merely a matter of desperation breeding a false optimism, a false hope?

Although we would not say that we are optimistic, we are hopeful. It is useful to review the situation a bit more to see why. Historically, in both the United States and in Italy there were movements of youth in the sixties that were even more interesting than the concurrent "war on poverty" in the former country. But those movements tended to take the form of a dropping-out of the institutional society, something that proved to be impossible, perhaps even by definition. Or they took the form, finally, of efforts to smash the institutions, a kind of nihilistic as well as politically terribly difficult alternative.

But from the maelstrom of the late sixties have come differing developments in the late seventies in the two countries. In Italy, the parties of the left, the Communist Party and the Socialist Party, have become orthodox proponents of reform and especially of the further development of a services society, albeit a more egalitarian one than that stressed by center and right parties. In addition, small ultraleft, so-called extraparliamentary fringe groups have multiplied among young people and students with diverse suggestions about the future. Apart from armed terrorist attacks their direct urban actions have tended to follow the partialized, sectoralized divisions of life forms and focuses that we have been criticizing. Housing-oriented actions have occurred: rent strikes, demonstrations and lobbying to obtain "fair rent" laws and actions by squatters occupying vacant housing. Actions occasionally have been directed toward moving local government to expropriate unused lands for parks.

In contrast, one finds in the United States a more innovative critique of the concept of services as well as of the institutional development of the service society. And it is not from the right that such critics as John McKnight of the prestigious Center for Urban Affairs at Northwestern University take on the core of the welfare state in their fight for new directions in urban policy making. McKnight's critique of services is in point and of a kind long overdue.

A potentially innovative development has occurred in Appalachia, a region in many ways even more depressed than the Italian Mezzogiorno, the south of Italy including Sicily. The notion of "community unions" has been born.[7] They focus on better living conditions, often emphasizing improvement of the usual range of educational, health, roads, welfare and local government services. But they are increasingly touching on the possibility of establishing or facilitating cooperatives, producer as well as consumer co-ops. The first co-op in the form of a coal mine owned cooperatively by a group of coal miners is but a few years old. Moreover, there are the first stirrings of such community unions toward efforts to capture local government control through the electoral process. These are, to repeat, potentially innovative and might constitute at least one of the ways in which urban planners might operate in a maieutic role in some American communities and regions whose urban future is most problematic.

With this we return to a point mentioned in Chapter Two, one that came as a surprise for this writer, an Italian urbanist: that the United States is not really a country with cities in crisis but a country that for a very long time has had very underdeveloped cities. Urban planners and urban specialists of all kinds have a special opportunity, indeed an obligation, to help to construct a meaningful alternative to the present minimal participation of people in minimally existing cities. The Yugoslav experience is relevant. We understand the relatively high participation there in a system of economic self-management as participation in specific places and spaces small enough to be understood as wholes, to some degrees at least, by their "citizens." It is an innovation supported by nearly all Yugoslavs, including most of those who strongly urge a return to a multiparty system.

The historical fact of the arrested development, indeed, premature decay, of American cities, is partially rooted in the fact that boundaries of cities there were never very meaningful for human experiences. Unlike the European city experience, suburbs and small towns outside of the historic centers of cities were always attractive to the classes dominant in America. For civic life for all citizens was substituted ethnic, neighborhood, and organizational life for fragmented and mobile populations. City boundaries were actually for the most part used for purposes of various demographic and population calculations or were otherwise formal lines drawn on maps, occasionally with city-limits signs posted at some of the points where thoroughfares crossed the boundaries.

A first giant step toward the construction of feasible systems of citizen participation, then, would be the initiation of projects, perhaps even calling for citizen conventions, to try to begin to create cities. The project would involve opening sets of institutions in particular localities and at the same time beginning to close cities, constructing meaningful boundaries around them. Such projects would involve some long-range imagining and policy-planning but with massive citizen participation. It would mean the creation of plans for dislocating structures in large areas that constitute much of the fabric of both suburbia and urban sprawl, but only after their inhabitants had moved or died. It would mean the beginning of the end of megalopolis, which is a contradiction in terms, and certainly not something that signifies mature cities.

It has been said that poorly planned or privately developed suburban developments have created an enduring social and physical fact. Sociophysical facts, we agree. Enduring? Not necessarily. Redevelopment and renewal are possible and possible without the kind of destruction of people and their neighborhoods that such policy programs have meant in the past. But we are not proposing instant transformations and certainly not those of the kind in the past or the present when people are forced out or removed by decisional processes in which they have had no real part. Nor do I think that our common humanity requires either homogeneous cities (or countries) or an unbounded sprawl; boundaries may represent selfish isolationism or companionably distinctive cultures.

A set of projects "to construct cities," not new towns nor new cities in the usual sense, is what I think the United States desperately needs. In Italy as in other European countries such projects would be more properly "rebuilding" or "reconstruction" or "revitalization" of cities. On both sides of the ocean this would mean great changes in the governmental system. It would mean in the United States beginning to take seriously the nation's long-cherished but until now superficial localism. It would mean in Europe a speeding up of the process of decentralization of government, as in Italy's new regional level. It could not mean in either place the creation of little, inefficient, isolated, city-centered, economic and other institutional nodes. It would mean a movement toward smaller-scale, more humanly satisfying, institutionally opened cities of civic participation.

It has often been said that the country has been glorified while the city has been vilified or at least distrusted, and especially by intellectuals.[8] The reader will understand that I have a procity bias although not of the usual kind. Today's cities in America are truly nightmares of chaos or boredom or loneliness or some combination of such conditions. Today's cities in Italy are quickly becoming Americanized. They are atrophying quickly and their future is as doubtful as that of cities in the United States. Of course I expect the double-barreled criticism that says I am glorifying European cities and proposing a romantic, unrealistic reactionary return to an impossible small-city America, long gone forever. Perhaps, but I doubt both parts of the criticism. What may happen in Appalachia, for example, is an entirely new kind of city, not a return to a colonial American kind of city culture. And what may be created in the way of cities in such places as New York, Chicago and Los Angeles I do not pretend to know. Mine is an argument for massive involvement of masses of people in a long process of city creation, re-creation and vitalization that, perforce, means redimensioning the entire array of the human arrangements that we term institutions and thus the present urban society itself.

It might be recalled that perhaps the first major step forward in modern times in opening decisional processes to participation by large numbers of people theretofore excluded, and still excluded in most other countries, came in Yugoslavia in the early fifties (and continuing to the present) especially in regard to self-management in industry and other economic domains. That was a time of stress for Yugoslav authorities who found themselves with a need to innovate and yet avoid either the Soviet central planning or the American private enterprise models, both extremely hierarchical. In the United States it is probably not too much to suggest that there is an approaching, if it is not here already, time of choice for planners and others, a choice between moving in a democratic direction or in the direction of increased dictation and ever less freedom in the urban society.

To support any increase in the control and coordination of the various dimensions of community life without an effective democratic presence of people as participants, means to move toward more dictatorial forms of control, however marked by scientific appearance or seeming efficiency. In modern urban society, people are increasingly partialized and, hence, constricted, distorted and dehumanized in the name of science or of efficiency. There is no justification for urban planners or other professionals assisting in further tightening the screws, even if there is nearly a consensus that the urban crisis is upon us and there is no other way to avoid the total breakdown of cities or the society itself. There *is* an alternative and the time for its beginning is now.[9]

The two objectives of participation and despecialization cannot be separated moments. They demand the combined commitment of administrators and urban planners. The first great task for administrators, experts and those students who are on the way to becoming future experts is to understand that an innovative transformation of society requires much of them as persons. They are already invested with roles according to the organizational forms shaped by advanced capitalist society. They must be among the first to go beyond these traditional roles even though they themselves act as if they were atomized specialists. They must be among the first new revolutionaries of the postmodern period. They must not be traditional revolutionaries trying to capture control of the existing institutional heights, nor avant-garde revolutionaries trying to abolish institutions mindlessly, nor mystical revolutionaries engaging only in transformations of consciousness, of minds only. Perhaps another word rather than "revolutionaries" should be found for the needed new role.

In this lies their chance to assume a more important pursuit than that of making present or reformed institutions function a little better. By going beyond their traditional to a maieutic role, they can immediately begin to transform institutions.[10] It is no longer a matter of waiting for the right moment for the revolution. And only through such a transformation can all of the contemporary subordinate classes and the millions of marginalized people, not only the financially poor, regain the capacity and competence that they have not had the chance to exercise as rights of universal human experience.[11]

Obviously, such an initial act of understanding is a first step in a process that begins but does not end with that moment. The fact of a progressive change in those who act in the organization of the city becomes increasingly different over the passage of time from a simple modification of some administrative processes or of the methods of a discipline such as urban planning.[12]

My hypothesis is that plans that are no longer produced according to traditional, abstract, standardized and fragmented models of urban man and the city do more than presume a less partialized and alienated subject. They in their turn influence, confirm and accelerate the process of such de-alienation and retotalization of more potent people.

NOTES

CHAPTER 1

1. Many observers or analysts of experiences in the United States, especially in community action programs, agree that initiatives have been blocked precisely because people proved that they could participate actively, make decisions, and, if anything, go beyond the narrow limits set by the government or the public authorities for such initiatives. See the work by Kenneth Clark and Jeannette Hopkins, *A Relevant War Against Poverty*, New York, Harper & Row, 1970, and Moffett Toby, *The Participation Put-on*, New York, Delacorte Press, 1971. That appeared to be the case also in 1948 as reported in one of the most remarkable social-science analyses of the methods of holding and wielding power in American cities, specifically in Atlanta, Georgia, termed therein Regional City, by Floyd Hunter, *Community Power Structure: A Study of Decision Makers*, University of North Carolina Press, 1953. We are not, however, asserting that under present conditions many of the poor or the marginalized are ready to participate under the general conditions of an unchanged system. See note 8, Chapter 1, *infra*.

2. The kind of plan produced influences people: "consumers" of the plans have or develop the same kind of understanding of reality that permeated and shaped the plans. Of relevance to this entire section of this Chapter is a critical survey of the American empirical literature on the matter of alienation and citizen participation by James D. Wright, who probes precisely the question of what might happen if the traditional nonparticipants began to participate. He is most optimistic and on the basis of some past studies showing differences in thinking and desires on the part of the participant haves and nonparticipant have-nots, he does not worry about the possibility of "co-optation" that we are discussing.

It is difficult if not impossible to use past studies for projecting a future of the kind we envisage, whereas Wright proposes much more citizen participation with traditional representational structures and thus can use past studies. Nonetheless his analysis is relevant to several of our major themes; see *The Dissent of the Governed: Alienation and Democracy in America*, New York, Academic Press, 1976.

3. In this regard the modern Italian experience offers a relevant and significant example. In Bologna, where the Communist Party has controlled the local government since the end of the war, one sees clearly that the aim of the Communists in managing the city has been to create the image of a "model city," testifying to their ability to manage a city in a way superior to that of their opponents. Within the scope of this aim citizens are encouraged to participate, but especially with the intention to enlarge consensus and support with respect to the Communist administration. For this reason the so-called "grass roots" initiatives have always had one thing in common: they were initiated and directed by members of the same Communist Party. When the party lost control of such initiatives in some outlying district committees where the Communist Party was not able to take control through its local members, the administration itself intervened in order to direct the action back to its initial paths.

4. From the historical economic perspective, Faenza flourished in two sectors: agriculture and craftsmanship. It was not ceramics alone that made it famous but also other sectors of the crafts such as ironwork, woodwork, leatherwork and the making of agricultural instruments. As is true in all of Italy, from the postwar period on the population began to abandon agriculture, but not to the extent it did in other areas. The development of the public service sectors, commerce and the creation of new industries absorbed laborers, especially those from the farms. If the economic events of Faenza are similar to those of many other municipalities of northern Italy, the political events are more particular. Indeed, although it belongs to the region of Emilia-Romagna where Bologna and almost all of the major municipalities had a Communist or Socialist-Communist majority since the immediate postwar period, Faenza has always had a large Christian Democratic presence, which controlled the local government until 1975.

5. Having stressed here that we do not pretend to give a detailed or even general image of an entire country through one or two communities, we must note nevertheless that other studies have illuminated the national representativeness of such community samples. Among them we can cite a comparative study sponsored by UNESCO in which samples of a community and of an entire nation were drawn and compared. One of the authors wrote: "If one were representing cultural differences with any fidelity in the time-budget distance data, one would certainly expect these part-whole relationships to show just such minimal 'distances.' By implication, furthermore, we claim some insurance that the selection of a particular city within a nation runs less risk of misrepresenting the nation as a whole than we might expect intuitively." (Philip E. Converse, Survey Research Center, the University of Michigan, "Gross Similarities and Differences in Time Allocations," p. 4, a progress report prepared for the VIth International Sociological Congress in Evian, France, September, 1966). On the same subject and on the small size of samples usually used in cross-national research, see also David C. McClelland, *The Achieving Society*, New York: A Free Press Paperback, Macmillan, 1967; E. Paul Torrance, *Rewarding Creative Behavior*, Englewood Cliffs, N.J., Prentice-Hall, 1965.

6. For a description of the beginnings of this project, see Robert E. Agger, Miroslav Disman, Zdravko Mlinar and Vladimir Sultanovic, "Education, General Personal Orientations, and Community Involvement," *Comparative Political Studies*, April 1971, pp. 90-116; a revised version appears in Jack Goldsmith and Gil Gunderson, eds., *Comparative Local Politics: A Systems-Function Approach*, Holbrook Press, 1973. The first findings of the major part of the study were reported by Disman, a member of the Department of Sociology of York University, in "The Values and Participation," *Participation and Self-Management*, Vol.

2 (First International Sociological Conference on Participation and Self-Management, The Institute for Social Research, University of Zagreb, 41000 Zagreb, Jezuitski Trg 4, 1972), pp. 44-72.

A more complete presentation and discussion of the international study is now being prepared by Agger, Disman and this writer. See also Disman and M. Ondracek, "Socio-economical Determinants of Participation in North American and East European Communities," a report prepared for the Annual Meeting of the Canadian Sociological and Anthropological Association, 1974; with respect to the two Italian communities, see Simona Ganassi, "La dimensione urbana dell'impegno civico: gestione del potere e partecipazione in due comuni dell'Italia settentrionale," a report to the National Research Council of Italy (C.N.R.), 1975.

7. Although there are often efforts made to rationalize such findings on the ground that low participation signifies a healthy political democracy (e.g., T. Dye and H. Zeigler, *The Irony of Democracy*, New York, Duxbury, 1972), for the most part the findings escape notice and come as much a surprise to American (and, indeed, foreign) political scientists as to American citizens of whatever educational level. That is true despite such empirical findings of comparative studies as reported in S. Verba and N. H. Nie, *Participation in America*, New York, Harper & Row, 1972, p. 340, and in James D. Wright, *op. cit.*

8. See the essays by R. E. Edwards, M. Reich and T. E. Weisskopf in the volume entitled *The Capitalist System*, Englewood Cliffs, N.J., Prentice-Hall, 1972, especially chapter 5, part III.

9. See J. Fiszman, *Revolution and Tradition in People's Poland: Education and Socialization*, Princeton University Press, New Jersey, 1972, and Pavel Machonin, *Czechoslovakian Society: A Sociological Analysis of Social Stratification*, Bratislava, 1969.

10. I do not mean to assert here that such sectors as those of university or of the artistic life are more "political" in Italy than in other countries; the distinction is between what pertains to "politics" and what to "political parties." Since party logic and the hands of parties are present in very many Italian institutions, the common belief is that adhesion to as well as involvement in parties is a general, or at least widespread, condition among citizens generally.

11. It is worth mentioning that from the postwar period to the present, the regions of the Veneto and Emilia-Romagna have had very different political characteristics. The first is the region of northern Italy where the Christian Democrats nearly everywhere had absolute electoral control. Emilia-Romagna, instead, has always been the strong point of the Communist Party. Not by chance was the Veneto called the Italian Vendée, the White region, and Emilia-Romagna the Red region.

12. The research report compiled for the National Research Council, which financed the Italian part of the project, tried to clarify the differences between the two Italian communities since, obviously, the comparison centered on them. In the international study—that is, in comparison with the other countries—it is clear that such differences have lost their relevance, being *comparatively* small in the cross-national community context.

13. Faced with the general marginalization of the elderly, it is especially interesting to note how in such a situation as Italy's there is a relatively great proportion of elderly in the power elites of the major institutions, especially the political and economic ones. In this regard journalistic language has produced a significant word for indicating the composition of those holding power in Italy: the "gerontocracy."

14. The text of the questions was the following:

Generally speaking, in the areas of community affairs in which you are most interested, do you think you could influence decisions if you wanted to? Yes; Sometimes yes and sometimes no; No (Don't know; No answer)

If you were concerned about a local community problem and contacted the appropriate officials, how do you think they would react? Which of the following statements best describes the way the officials would respond to you?
They would understand my problem and do what they could about it;
They would listen to me but try to avoid doing anything;
They would ignore me or dismiss me as soon as they could.
(Don't know; No answer)

15. For a wide variety of related research, see Herbert M. Lefcourt, *Locus of Control*, New York, Halstead Press of John Wiley, 1976. The items on the international questionnaire showing strong interrelations and within a single factor domain were as follows:

1) It's not worth worrying; what will happen will happen.
2) I prefer to do things that do not require much thinking.
3) I don't feel very self-confident when I have to talk with people I don't know very well.
4) It is a waste of time to pay attention to somebody with whom you disagree completely.
5) I doubt that men will ever solve the major problems of the world.
6) I would not even try to talk to someone who has completely opposite opinions.
7) I do not like to think much about complicated ideas.
8) I really cannot have as a close friend a person who does not share my beliefs.
9) You can't change human nature.
10) I do not usually like to start conversations because I don't feel I can speak very well.
11) The world is governed by supernatural forces which determine the course of events.
12) I prefer to have a simpler rather than a more complicated job.
13) I am worried about speaking publicly because I don't know how to express myself very well.
14) I feel uncomfortable talking in a group of people.
15) If no one would agree with some idea of mine it is better for me to forget it.
16) There is very little I can do to change what life has in store for me.
17) I prefer to do the same tasks in my work every day.
18) I sometimes worry about being laughed at when I speak to people.
19) I frequently feel it is my duty to agree with the opinions of my friends.
20) I would rather do familiar tasks than always to face new problems.
21) I feel uncomfortable when my opinions differ from others.
22) I would rather change my opinions than lose my friends.
23) I'd rather not solve my problems if the only way to solve them is to make enemies.
24) I do not like to have close contacts with people who have different leisure-time or recreational interests than I do.

The response categories offered were: strongly agree; somewhat agree; somewhat disagree; strongly disagree (no answer). Statements 8, 14 and 21 were also used in the short-form scale in Faenza; statement 20 was used only in the short-form scale in Faenza.

16. With regard to this subject Robert E. Agger has presented a suggestive new approach in a paper at a Yugoslav UNESCO commission conference on "Adult Education, Involvement and Social Change" held in Ljubljana, Yugoslavia in May 1970. The title of his paper was "A Cross-national Community Study of Civic Involvement: Some Empirical Findings and Notes Toward a Theory" (Institute of Sociology, University of Ljubljana, Cankarjeva 1, Ljubljana).

CHAPTER 2

1. A most thorough contribution to this very argument is in the first chapter of David Harvey, *Social Justice and the City*, London, Arnold, 1973.

2. This is an expression used by Henri Lefebvre in *The Urban Revolution (La Révolution Urbaine)*, Paris, Gallimard, 1970, p. 42.

3. For analysis of relations between economic and urban expansion during the first industrial epoch see Lewis Mumford, *The City in History*, New York, Harcourt, Brace, 1961, chapter 14.

4. Thomas Blair, *The International Urban Crisis*, London, Paladin, 1973, chapter 1.

5. For a good analysis of European multinationals, see Anthony Sampson, *The New Europeans*, London, Panther Books, 1964.

6. For a case study of this process in American society and some consequences, see A. J. Vidich and J. Bensman, *Small Town in Mass Society*, Princeton, N.J., Princeton University Press, 1968.

7. Lefebvre, *op. cit.*, p. 23.

8. Harry Braverman describes the process, beginning with so-called Taylorism, in the context not only of the organization of labor but also of "clerical" workers in offices, through some of the effects of the introduction of computers. See his *Labor and Monopoly Capital: The Degradation of Work in the Twentieth Century*, New York, Monthly Review Press, 1974.

9. Many books deal with this transition. See Mumford, *op. cit.*; see also C. Aymonino, *Origini e Sviluppo della Città Moderna*, Padua, Marsilio, 1971.

10. Le Corbusier, *The Athens Charter*, New York, Grossman, 1973, pp. 95-6. See Point 77 and Point 78; in fact the entire document is worth reading.

11. For a penetrating critique of zoning principles and various examples of and suggestions for cities and towns thought "ideal" in such terms, see Christopher Alexander, "A City Is Not a Tree," *Architectural Forum*, Parts I and II, 122 (April/May 1965), pp. 58-62. See also G. De Carlo, *Questioni di Architettura e Urbanistica*, Urbino, Italy, Argalia Press, 1961.

12. Harvey Lithwick, the chief designer of the Canadian federal ministry of urban affairs, raised the possibility of affording people choices of different "kinds" of cities through purposeful public policy. He attributed the suggestion to implications of Kevin Lynch's findings that different cities seem to have different images in the eyes of citizens in the United States. See Lithwick, *Urban Canada: Problems and Prospects*, Ottawa, Central Mortgage and Housing Corporation, 1969. This is Roberto Guiducci's thesis in his *La Città dei Cittadini*, Milan, Rizzoli, 1975. Guiducci holds that by situating various specializations in particular urban places, one eliminates the competitive situation that now exists among cities.

13. The notion of action-reaction with respect to territory and its control by the dominant class is developed by Magnaghi *et al.* in *La Citta' Fabbrica*, Milan, Clup Press, 1971. They suggest that when forced to undergo a certain degree of power-sharing by labor unions and workers in one phase of the class struggle, the dominant class recouped by elaborating new forms of exploitation of land in a subsequent phase.

14. For an analysis of consumer society from a Marxist, so-called Frankfurt School or critical-theory perspective, see Herbert Marcuse, *One-Dimensional Man*, Boston, Beacon, 1964. For a more straightforward Marxist approach, see J. Baudrillard, *La Societa de Consumi*, Bologna, Il Mulino, 1976. For an interesting non-Marxist analysis by an urbanist also critical of urban planning, see R. Goodman, *After the Planners*, New York, Simon and Schuster, 1971. In a very extensive literature, one might see also C. Wright Mills, *The Power Elite*, New York, Oxford University Press, 1957.

15. Lefebvre, *op. cit.*, p. 125.

16. Marginalization and the marginalized are terms used throughout this book to refer to these people on the fringes, on the margins, of society, who are regarded by those who share the major cultural perspectives of the particular society almost as aliens, as alien-ated, but not quite. They are more usually regarded, when other people are forced to regard them, as useless and potentially bothersome and annoying unless they remain quiet, faceless, anonymous, in their essentially secondary, nonproductive roles. They may become even more than annoying if they actually make demands for special treatment or services.

In the modern, advanced industrial society, at least in most and certainly in ours, there are also people with comparatively little or no economic or political wealth and power. We use the term "subordinate class(es)" or similar terms for such people. These include various groups of production or factory workers as well as many kinds of white-collar workers. These categories, the marginalized and the subordinated overlap. Some people or strata may be in both categories. Also, people may move over the passage of time into one and then, while remaining or leaving that category, into the other. That depends on the course of historical events.

We are especially concerned in this book with marginalized groups, such as the elderly and women generally, as well as housewives particularly, because they are so often slighted in traditional class or social analysis. Housewives had, in earlier, more rural and agricultural society times, been even more subordinated but less marginalized. Now, when not holding jobs, they tend to be more frequently only the latter. The "wages for housework" movement in the United States and abroad is an effort, as a first step, to return housewives to a less marginalized although still subordinated position, but not more subordinated than counterpart males.

Increasing numbers of unskilled workers, or potential (because young) unskilled workers are finding themselves increasingly marginalized. They are not (even) subordinated because they are without any work or much chance of finding work. They are increasingly not even secondary but totally surplus, and even parasitic, in the eyes of many of the economically productive. Black Americans, increasingly in such situations, are thus increasingly marginalized as well as racially rejected and, when working, often exploited much as women of both races still are. See S. Willhelm, *Who Needs the Negro?*, Garden City, N.Y., Doubleday Anchor Books, 1971.

17. See the classic and still relevant work by F. Engels, *The Housing Question* in Karl Marx and Friedrich Engels, *Selected Works*, Moscow, Progress Publishers,

1969, Vol. II. See also the bibliographically rich work: F. Indovina, ed., *Lo Spreco Edilizio*, Padua, Italy, Marsilio, 1972.

18. Lewis Mumford, *The Highway and the City*, London, Secker and Warburg, 1964.

19. F. E. Emery, ed., *Systems Thinking*, Harmondsworth, Penguin, 1969; R. L. Ackoff, *Scientific Method*, New York, John Wiley, 1968; C. W. Churchman, *The Systems Approach*, New York, Delacorte Press, 1968; and see the critical review and excellent bibliography by Francis Ferguson, *Architecture, Cities and the Systems Approach*, New York, Braziller, 1975.

20. This subject, that is, the possibility of conducting "objective" and "value-free" operations, has generated an interesting debate in Marxist circles recently and has entered thereby the Italian and European popular press. See, for examples of the American literature, such articles as C. Bay, "Politics and Pseudo-politics," *The American Political Science Review*, LIX, no. 1 (March 1965), pp. 39-51. That was written in reference to political science; there are few such treatments in regard to similar scientific currents in planning or architecture. Exceptions include (and others are noted in) William F. Hornick and Gordon A. Enk, "Analytical Paper on Human Values in Technology Assessment," The Institute on Man and Science, Rensselaerville, N.Y. 12147. For a book both interesting and relevant to the latter although written in most general terms, see C. W. Churchman, *Challenge to Reason*, New York, McGraw-Hill, 1968.

21. J. B. McLaughlin, *Urban and Regional Planning*, London, Faber and Faber, 1971.

22. This raises the question of whether people should or can renounce intuition. In its classical form the problem was that of producing solutions to urban problems in which only the expert was authorized to express such solutions. But only in the case of artistic creation were such ideas clearly understood as subjective. With regard to urban planning, one always resorted to the screen of "technical" values deriving from the knowledge, ability and rationality of the specialists, the experts, in order to justify their monopoly in that sector. Concerning intuition and its role in so-called "scientific creativity," see Y. Dror, *Public Policy Making Re-examined*, Chandler, San Francisco, 1968; and Churchman, *op. cit.*

23. Fuller, *Utopia or Oblivion: The Prospects for Humanity*, New York, Bantam Books, 1969; Doxiadis, *Ekistics, An Introduction to the Science of Human Settlement*, London, Oxford University Press, 1960; and his *Architecture in Transition*, London, Oxford University Press, 1963.

24. A. Touraine, *The Post-Industrial Society*, New York, Random House, 1971; E. Trist, "Urban North America, The Challenge of the Next Thirty Years," in *Plan*, the journal of the Town Planning Institute of Canada, vol. 10, no. 3 (1970), pp. 4-19.

25. The Club of Rome is a confirmation of this; we can call it a multinational group of brains that brings together technocrats from various capitalistic countries. To implement their proposals, they are able to get the highest political authorities of the countries interested to participate in their meetings. It is in connection with the Club of Rome that Jay Forrester produced a model of worldwide dynamics published in his book *World Dynamics*, Cambridge, Mass., Wright-Allen Press, 1971. It is not by chance that Forrester is concerned with global dynamics; he constructed the first so-called dynamics computer-simulation model of a city, illustrated in detail in his book *Urban Dynamics*, Cambridge, Mass., MIT Press, 1970. See the critique of the latter in Ferguson, *op. cit.*, pp. 57-58 and in Martin Kuenzlen, *Playing Urban Games: The Systems Approach to Planning*, Boston, The I Press, 1972.

26. On various occasions high officials of world finance such as David Rocke-feller and Robert McNamara declared that there would no longer be any large fi-nancial project or help given to underdeveloped countries not based on systems-analysis evaluation of the objectives and methods of such a request for aid. Also, the Food and Agricultural Organization of the UN decided to carry out only projects conceived as based on systems analysis.

27. M. Webber, "Culture Territoriality and the Elastic Mile," a paper of the Regional Science Association, 1964.

28. In *Social Justice and the City* Harvey divides his book into two parts. The first part he wrote during a period in which he had a "liberal" ideology. In the second part he confronts the same problems according to a Marxist reading of society. In the first part he upholds the thesis that the urban system proceeds to-ward "cultural heterogeneity and territorial differentiation" (p. 84). As is clear from my position here, I do not share this thesis. Harvey does not explicitly re-turn to this concept in the second part.

29. This is of course not true for all sectors yet, as is testified to by residential housing that is still largely in private hands and is still a source of huge specula-tive profits.

30. Karl Marx, *Capital*, Moscow, Foreign Languages Publishing House, Vol. I, part IV, chapter XIV. That part is especially concerned with the division of labor in the factory as well as in society. See also G. Stevenson, "The Social Relations of Production and Consumption in the Human Services," *Monthly Review*, no. 9 (September 1976).

31. See the special number edited by Roystan Landau, "The Individual in an Institutionalized World," *Architectural Design* February 1976. Among the vo-luminous critical writings on institutions generally and by type, see the works of D. G. Cooper of the Tavistock Institute on the family and on the "dialectics of liberation"; by Ivan Illich on schools, on medicine and his earlier "call for insti-tutional revolution"; and the works of the Italian critical psychiatrist F. Basaglia: *L'istituzione Negata*, Turin, Einaudi, 1968; and *La Maggioranza Deviante: L'Ide-ologia del Controllo Sociale Totale*, Turin, Nuovo Politecnico, Einaudi, 1971. The latter has an international institute for the criticism of institutions, located in Venice, Italy.

32. Ernest Erber, ed., *Urban Planning in Transition* (for the American Insti-tute of Planners), New York, Grossman 1970, p. xvii.

33. R. L. Warren, *Studying Your Community*, New York, Free Press, 1965, p. 61.

34. Herbert Gans, "The City and the Poor: The Shortcomings of Present Housing and Anti-Poverty Programs, and Some Alternative Proposals for Elim-inating Slums and Poverty" in Paul Meadows, ed., *Urbanism, Urbanization and Change: Comparative Perspectives*, Reading, Mass., Addison-Wesley, 1969.

35. We hope that with this statement we are not in the tradition of what Mor-ton and Lucia White term "organic metaphysics." They charge Lewis Mumford with having such a perspective; of having "blurry blueprints." We agree with what we understand to be Mumford's central message although we put it quite differently from the way he did in the passage quoted by the Whites. Indeed, our blueprints are blurry, but our message is that construction as well as reconstruc-tion of cities must no longer be attempted via elitist architectural or urban plan-ning, or economically dominant corporate or government power but by opening the process of designing cities to masses of their citizens. See the Whites' *The In-tellectual Versus the City*, Cambridge, Mass., Harvard University Press & MIT Press, 1962, p. 236.

36. See Sam B. Warner, Jr., *The Private City*, Philadelphia, University of Pennsylvania Press, 1968.

37. We recognize the validity of Don Martindale's point that with the national level of importance of government and business and other matters the city is no longer a community for many: "the new community [is] represented by the nation." We do not agree with him that taking up arms to defend the city's walls, thus making the city a unit of survival to which people owe their supreme loyalties, was the enduring, necessary element in the existence of cities in all past history nor that it must be in future. He infers that from Max Weber's theory of a city, but we find the latter's approach a bit too "institutional" and not sufficiently social psychological, to use Martindale's own categories distinguishing various approaches to the conception of the city. See Max Weber, *The City*, "Prefatory Remarks" by Don Martindale, Glencoe, Ill., Free Press, 1958, pp. 55-62.

38. We refer to Gabriel A. Almond and Sidney Verba's *The Civic Culture*, Boston: Little, Brown, 1965.

CHAPTER 3

1. Erber's definition of city planning is rather typical: "City planning was the only area of American life and letters concerned with the sytematic examination of the techniques of purposeful and systematic change of the natural and man-made environment." See *Urban Planning in Transition*, New York, Grossman, 1970, p. xviii.

2. G. DeCarlo, "L'architettura della Partecipazione" in *L'architettura degli Anni 70*, Milan, Il Saggiatore, 1973.

3. See the preface written by L. Quaroni for M. Cerasi's *Lo Spazio Collettivo della Città*, Milan, Mazzotta, 1976.

4. M. Cerasi, *op. cit.*

5. Erber, *op. cit.*

6. Frances Fox Piven, "Planning and Class Interests," *AIP Journal*, no. 41 (September 1975), p. 308.

7. *Subject to Approval*, a useful comprehensive and critical review of municipal planning in Ontario, published by the Ontario Economic Council (Queen's Park, Toronto), edited by Comay Planning Consultants, Ltd., and P. S. Ross & Partners, 1976.

8. In my opinion this problem is based on pseudoproblems and not only applies to urban planning and architecture-urban planning. In many other disciplines similar perspectives have been debated and are still being debated. In the discipline of sociology, for example, one finds absurd discussions between urban, general, and systems sociologists, for example, about what belongs properly to each "field." But the discussions become even more absurd when they become efforts to cut in two such "fields" into "theoretical" and "applied" domains, or the discipline of sociology itself into social theory and other things.

9. David Harvey, *Social Justice and the City*, London, Arnold, 1973, p. 303.

10. Lewis Mumford, *The City in History*, New York, Harcourt, Brace 1961, p. 421 *et passim*.

11. Apart from the differences existing in the urban planning legislation of various nations, this constitutes a constant in both capitalistic and socialist countries. In other words, in its essence the plan is still everywhere constituted by prescriptions visualized graphically with symbols on topographical maps.

12. Frances Fox Piven points out that American planners of the fifties were,

from their position within the university, ready to confront urban organization without ever verifying what social strata their proposals were actually benefiting or damaging. They took it for granted that whatever helped the city to expand was *a priori* fair and in the general interest. It has already been demonstrated and is well known that the choices of the plan in the context of capitalistic society are not neutral. My thesis here, however, goes beyond that point to confront the problem of whether or not there are any opportunities for plans as they are presently conceived to be other than unjust and alienating even in socialist countries.

13. Stafford Beer, *Designing Freedom*, Toronto, CBC Publications, 1974, p. 8. The work of W. R. Ashby was crucial for Beer.

14. This fact has been repeated so frequently that it is not necessary to give specific references. Every city presents examples. Then there are those huge "urban renewal" operations in North America where, in the name of "rehabilitating" decaying areas, crowded and socially alive quarters are swept away. Herbert Gans has documented this with respect to Boston in his book *The Urban Villagers*, New York, Free Press, 1962. In 1967 the municipality of Montreal also carried out studies on the quarter of Victoria, wanting to show the extreme degree of social degradation there as well as a public interest in remedying it by building a new sports complex. According to accurate "sociological" and "urban planning" studies, this primarily Italian-Canadian quarter proved to have a high degree of social vitality and good social qualities, with little criminality or alcoholism. Despite that, the quarter was swept away. As in other instances, the inhabitants were given help, however inadequately, in their attempts to relocate themselves elsewhere in the suburbs of the metropolises of Montreal which were then in the process of full expansion. But the relocation was also a diaspora, a dispersal of a relatively alive community.

15. I am referring here to the so-called "C.I.T.A. Struggle" in the marginal area between Marghera and Mestre, where the Venetian mainland begins, but other examples are numerous.

16. I am referring here to the well-known events of the installation of the E.N.I. refinery in Lugugnana near Portogruaro in the province of Venice.

17. These concepts are developed in Henri Lefebvre's *La vie quotidienne dans le monde moderne*, Paris, Gallimard, 1969; see also B. Brown's *Marx, Freud and the Critique of Everyday Life: Toward a Permanent Cultural Revolution*, New York and London, Monthly Review Press, 1973.

18. See the criteria set forth in the introduction to the collection of essays by R. Mainardi, *Le grandi città italiane: Saggi geografici e urbanistici*, Milan, F. Angeli, 1971.

19. Despite this, such parameters are often used as examples in preparing the laws that define the areas in which special aid is granted or public intervention is programmed.

20. Beer in *Platform for Change*, New Jersey, John Wiley, 1974, refers to his experiences in this direction that he had in Chile during the Allende years. He was directly commissioned by Allende and consulted with him personally.

21. See many relevant works: L. Srole, *Mental Health in the Metropolis*, New York, McGraw-Hill, 1962; Philip Slater, *The Pursuit of Loneliness: American Culture at the Breaking Point*, Boston, Beacon, 1976; R. Sennet, *The Uses of Disorder*, London, Allen Lane, Penguin Press, 1970; A. C. Rennie *et al.*, "Urban Life and Mental Health: Socioeconomic Status and Mental Disorder in the Metropolis," *American Journal of Psychiatry*, 113: 831, 1957; R. E. L. Faris and H. W. Dunham, *Mental Disorders in Urban Areas: An Ecological Study of Schizophrenia and Other Psychoses*, Chicago, University of Chicago Press, 1939; D. C.

Glass and J. E. Singer, *Urban Stress*, New York, Academic Press, 1972.

22. Karl Marx, *Grundrisse*, The Pelican Marx Library, trans. by M. Nicolaus, Harmondsworth, Penguin, 1973, p. 92.

23. A. Cuzzer, "Compendio dello studio sui temi urbanistici," in *Il Governo delle città*, F. Fiorelli, ed., Milan, F. Angeli, 1975.

24. L. Jakobson, "Toward a Pluralistic Ideology in Planning Education" in Erber, *op. cit.*, p. 271.

25. In the article cited earlier "L'architettura della Partecipazione," G. DeCarlo explicitly refers to such a vision of the city. It is this philosophy that marks the entire process leading to the unchallenged supremacy of the methodology of zoning in modern planning.

26. C. Aymonino, *Il significato della città*, Bari, Laterza, Italy, 1973.

27. Such principles have been and continue to be references for Soviet urban planners as well. M. V. Posokhin is the chief architect of Moscow and has authority in urban planning and politics as well, having been elected a representative to the Soviet of Moscow and to the Supreme Soviet of the USSR. In his book *The City To Live In*, Moscow, Novosti Press Agency Publishing House, 1974, Posokhin continuously confirms the validity of the principles of so-called rational planning. He asserts (p. 12): "Le Corbusier's *La Charte D'Athenes* on the whole correctly describes the totality of tasks involved in achieving balanced urban development." Although their objectives are quite different, urban planners in Eastern Europe generally operate similarly to American and Western European planners methodologically as well as sharing in large part the model of the "good" city. They are led by the dictates of their profession to some of the same kinds of problems that American urban planners face or create as, for example, in whose interests or with what priorities are uses to be assigned to various kinds of land uses as well as public services including state subsidized housing. That is made clear in a survey by Ivan Szelenyi, "Urban Sociology and Community Studies in Eastern Europe," *Comparative Urban Research*, vol. IV, nos. 2 and 3 (1977), pp. 11-20. See also his references to the situation in the Soviet Union therein.

28. Every program involving the making of the plan now begins with the declared goal of making a participatory plan. In Chapter One we examined what this actually means in practice for urban planning. In the most recent writings of a very long list concerned with this issue, we find the dominant note to be a repetition of the same hortatory concepts but without any contribution to solving any of the real problems that such a perspective involves. In this regard see also the writings in P. Guidicini, ed., *Gestione della città e partecipazione popolare*, Milan, F. Angeli, 1973. Or see the book by the professional planner called by Nathan Glazer "the best social planner of the 1960s," Marshall Kaplan, *Urban Planning in the 1960s: A Design for Irrelevancy*," Cambridge, Mass., MIT Press, 1974.

29. A. Touraine refers to "dependent participation" in his *Post-Industrial Society*, New York, Random House, 1971. Among other authors of general sociopolitical analyses and of empirical studies who come to much the same conclusions as we do, see N. Dennis, *People and Planning*, London, Faber and Faber, 1970; A. J. Vidich and J. Bensman, *Small Town in Mass Society*, Princeton, N.J., Princeton University Press, 1968; J. Sewell, *Up Against City Hall*, Toronto, James, Lewis & Samuel, 1972; W. A. Gamson, *Power and Discontent*, Homewood, Ill., Dorsey Press, 1968, especially pp. 139-142; and Brian J. Heraud, "The Development of Community in New Towns" (Polytechnic of North London, Department of Sociology, unpublished paper, 1976).

30. "Utopia, The City and The Machine," *Daedalus*, Spring 1965, p. 290.

31. Marx says in *Grundrisse*; "Consumption produces production. . .because a product becomes a real product only by being consumed. For example, a garment becomes a real garment only in the act of being worn; a house where no one lives is in fact not a real house; thus the product, unlike a more natural object, proves itself to be, *becomes*, a product only through consumption. . .If it is clear that production offers consumption its external object, it is therefore equally clear that consumption *ideally posits* the object of production as an internal image, as a need, as a drive and as purpose. It creates the objects of production in a still subjective form. No production without a need. But consumption reproduces the need." *Op. cit.*, pp. 91-2.

32. This hypothesis is sufficiently dealt with by E. F. Schumacher in his *Small Is Beautiful*, New York, Harper & Row, 1973.

33. See the works by Schumacher and Beer already cited. Among Americans, David Vail of the Department of Economics, Bowdoin College, has done important studies on the possible use of alternative or intermediate technologies.

34. L. Jakobson in "Toward a Pluralistic Ideology in Planning Education" in Erber, *op. cit.*, comes to this conclusion (p. 275) as does Thomas D. Galloway and R. G. Mahayni in "Planning Theory in Retrospect: The Process of Paradigm Change," *AIP Journal*, January 1977, p. 68.

35. G. Astengo, "Per una pianificazione operativa" in *Gestione del territorio*, a special number of *L'architetto* of the National Council of Architects, February 1976, p. 41.

36. This is supposedly the case in countries as different as Italy and the United States. From the perspective of the formation of plans and in the light of the bureaucratic procedures used in adopting them, the situations of the Western nations at least are not substantially different, even if the names of the institutions and procedures involved in these operations obviously vary.

37. M. Meyerson, "The Next Challenge for the Urban Planner," *AIP Journal*, January 1977, pp. 371-76. A cohesive antiplanning force is to be found among members of the Public Choice Society and in the Institute for Contemporary Studies. See the latter's direct confrontation with the Initiative Committee for National Economic Planning in the sponsorship of the proposed legislation to provide for "Balanced Growth and Economic Planning" (1975) in *The Politics of Planning: A Review and Critique of Centralized Economic Planning*, Institute for Contemporary Studies, 260 California Street, Suite 811, San Francisco, California, 1976.

38. R. A. Cohen, "Small Town Revitalization Planning: Case Studies and a Critique," *AIP Journal*, January 1977, p. 9.

39. See Robert L. Bish and Vincent Ostrom, *Understanding Urban Government*, Washington, D.C., American Enterprise Institute for Public Policy Research, 1973, especially pp. 7-10, for an opposite point of view.

40. H. Gans, "Planning for Declining and Poor Cities," in *AIP Journal*, no. 41 (September 1975), p. 305.

41. P. Davidoff, "Working Toward Redistributive Justice," *AIP Journal*, no. 41 (September 1975), pp. 317-18.

42. Norman Krumholtz, Janice M. Cogger and John H. Linner, "The Cleveland Policy Planning Report," *AIP Journal*, no. 41 (September 1975), p. 299.

43. *Ibid.*, p. 298.

44. Piven, *op. cit.*, p. 310.

45. Krumholtz, Cogger and Linner, *op. cit.*, p. 299.

46. Kaplan, *op. cit.*, emphasis added.

47. "Who Can Activate the Poor?" by Warner Bloomberg, Jr., and Florence W. Rosenstock, in *Community Politics*, ed. by Charles M. Bonjean, T. N. Clark and R. L. Lineberry, New York, Free Press, 1971, p. 151.

48. *Ibid.*, p. 152.

49. Kenneth Clark and Jeannette Hopkins, *A Relevant War Against Poverty*, New York, Harper & Row, 1970, pp. v, vi.

51. *Ibid.*, pp. 27 ff.

52. The saddest position, but also popular with many planners since it is the easiest for them to work with, is one of neglecting the poor in order to induce, somehow, mobility into the middle classes. Perhaps the most notable example of that in recent years is the book by Edward C. Banfield, *The Unheavenly City*, Boston, Little, Brown, 1968. That work seemed to strike a responsive chord among many urban planners, among others. See the devastating criticism in the review of Banfield's book by Robert E. Agger, *Social Science Quarterly*, Spring 1971, pp. 835-53.

CHAPTER 4

1. I do not intend by this term the "lumpenproletariat" nor only the proletariat. See the interesting cover story with the former definition in "The American Underclass," *Time Magazine*, August 29, 1977. I use the terms underclass, subordinate classes, the dominated and the exploited interchangeably in the book. I refer to social strata constituted not only by the traditional working classes, blue-collar workers, but also by those who in earlier eras were part of distinctive middle-class strata, part of the lower bourgeoisie. Then those white-collar workers were in intermediate positions between the dominant directing class and the proletariat; today these strata have become massified without any real decisional power. For a study of the process of massification, especially in regard to clerical work, see Harry Braverman's *Labor and Monopoly Capital: The Degradation of Work in the Twentieth Century*, New York, Monthly Review Press, 1974. Another study in point is the massive landmark Eastern European empirical research headed by Pavel Machonin, with an English summary by M. Petrusek, in *Czechoslovakian Society: A Sociological Analysis of Social Stratification*, Bratislava, 1969.

2. Various analysts tend to give increasing importance to the lack of imagination among politicians and bureaucrats and cite this as a factor affecting the growing incapacity of modern governments to govern. There seem to be, however, some examples of innovation from the Third World, at least with regard to alternative ways to meet urban problems. Richard Martin's "Institutional Involvement in Squatter Settlements," in the English review *Architectural Design*, April 1976, gives an interesting example of such innovations.

3. I am particularly referring here to psychiatrists, psychoanalysts, psychologists, and so forth. In a country like the United States the use of their "services" may become more than a "privilege" of the well-to-do classes who can afford the high costs of such services; they may become the "right" of everyone because the state may make them available to everyone, given the enormous advantage such services offer in terms of political stability. One can see how widespread the "use" of these specialists now is from the fact that remarks about going to one's analyst are very common not only in novels and films but also in cartoons appearing in the daily newspapers. Even young children are sent by the affluent to analysts in the United States, particularly where analysts are available.

4. Stafford Beer, *Cybernetics and Management*, London, 1959; also by the same author, *Decision and Control*, London, 1966. See the other works of Beer cited earlier.

5. There are few writers who decisively favor demystifying the superhuman qualities of institutions although many come close to doing so. Most carry out their critical analyses with the intention of *improving* the functioning of institutions. See, for example, the analyses of Peter Berger and Thomas Luckman in their book *The Social Construction of Reality*, Garden City, N.Y., Doubleday, 1966.

6. Herbert Marcuse, *An Essay on Liberation*, Boston, Beacon Press, 1969; and the same author's *Five Lectures*, Boston, Beacon Press, 1970.

7. The most interesting definition in this respect comes from the work of Henri Lefebvre, especially from the analysis he made in his *La révolution urbaine*, Paris, Gallimard, 1970. See also David Harvey *Social Justice and the City*, London, Arnold, 1973, chapter VI, "Urbanism and the City: An Interpretative Essay," and pp. 130-35 wherein he examines critically the perspectives of the Chicago school of urbanists.

8. M. Casteils, *La questione urbana*, Padua, Marsilio, 1974, p. 109.

9. See the insightful approach by Raymond Williams in *The Country and the City*, London, Paladin, 1975. In the final chapter he stresses appropriately, although not at great length, that attention should be given to the organization of human life and transforming the division of labor, in a spirit close to our approach.

10. *The New York Times*, reported in *The Globe and Mail*, Toronto, April 13, 1976.

11. Such criteria are drawn from such works as A. Maslow, *Motivation and Personality*, New York, Harper & Row, 1954 and *Toward a Psychology of Being*, New York, Van Nostrand, 1968; H. A. Murray *et al.*, *Explorations In Personality*, New York, Oxford University Press, 1938; Richard de Charms, *Personal Causation*, New York, Academic Press, 1968.

12. See the study by B. R. Clark, *Educating the Expert Society*, San Francisco, Chandler, 1962. See also the writings of such political sociologists as Floyd Hunter, C. Wright Mills and William Domhoff. One of the most thorough treatments of this subject can be found in the collection of essays *Scuola, Potere e Ideologia*, Bologna, Il Mulino, 1972.

13. I am referring particularly to such writers as B. F. Skinner and his utopian vision described in *Walden Two*, New York, Macmillan, 1948. Skinner's error is even more profound. He has a model of man as totally determined, as a set of effects of outside forces. See his *Beyond Freedom and Dignity*, New York, Knopf, 1971. See also the general issue on "Utopia" in *Daedalus*, Spring 1965, especially the article by Lewis Mumford, "Utopia, The City and The Machine," pp. 271-292. Finally, and somewhat sadly, I note the kind of "utopia" proposed or anticipated by Percival Goodman some thirty years after his classic study of community planning, *Communitas*, with his brother Paul; the later volume is entitled *The Double E*, Garden City, N.Y., Doubleday Anchor, 1977.

14. Such a reading is also consistent with important parts of the works of Marx. Particularly to the point is the first part (with Engels) of *The German Ideology*, C. J. Arthur, ed., New York, International Publishers, 1970, pp. 37-95, as well as *Grundrisse* as others. See also the work on the nonexistence of institutions and ending the divisions of life and labor by Robert E. Agger, *A Little White Lie*, New York, Elsevier, 1978.

15. One should note here that the concept "to represent" is also rooted in those who fight to change society. Herbert Gans is a well-known American urban

sociologist who views the urban situation from a relatively radical perspective. He has made devastating critiques of urban planning politics of the sixties typical of most cities in the United States (for example, urban renewal). See his *People and Plans*, New York, Basic Books, 1968. Yet Gans evaluated the new plan of the city of Cleveland in a highly positive way. The major reason for such a judgment, apart from the bleak picture of the relations of the planners to the poor in other cities, is that the Cleveland urban planners decided to side with the poor and to make a plan that would best represent their interests. In the Cleveland Policy Planning Report one sees how the usual considerations are still preoccupying: a certain emphasis on public transportation, a program for public housing, programs for public services. All this, however, is very traditional and, we anticipate unhappily, is destined also traditionally to remain on paper. No new possibilities are opened concretely to the poor "for whom" planning is now to be done, but not in such a way as to afford them more decisional potency, more chances to participate and to express themselves in a planning process that would be done with them rather than for—or against—them.

16. We do not need to underline the point that this is not only true for urban planners. B. Bernstein in his book *Class, Codes and Control*, St. Albans, Herts, (Paladin, 1973), examines the relations between the codes of specialized language and the class organization of modern society especially but not only in regard to schools. For a prototypical description of what happens in the United States when people are threatened by a highway relocation and some try to protest, in this case to highway planners, see Gordan Fellman, "Neighborhood Protest of an Urban Highway," *AIP Journal*, March 1969, pp. 118-122, and Fellman, Brandt and Rosenblatt, "Dagger in the Heart of Town," *Transaction*, September 1970, pp. 39-47. In addition to their comments on the "experts," they have some useful analysis of what kinds of trivializing and dehumanizing psychodynamics are experienced by the people who appear to protest in this ritual of citizen participation. At that time, 1970, Fellman *et al.* suggested what proved later to be an unsuccessful reform for ameliorating the situation, a new role termed advocacy planners (although they did not use that name). See the latter article, p. 47.

17. See the set of insightful analyses by Theodore Roszak: *The Making of a Counter Culture: Reflections on the Technocratic Society and Its Youthful Opposition*, Garden City, N.Y., Doubleday Anchor, 1969 and *Where the Wasteland Ends*, Garden City, N.Y., Doubleday, 1972.

18. The process is described more or less in this way by many urban planners, among whom is the aforementioned A. Cuzzer, in *Il Governo delle città*, ed. by F. Fiorelli, Milan, F. Angeli, 1975, p. 29.

19. There are many writings that deal with the concept of the "model." One of the most complete and clear contributions is that found in C. W. Churchman and R. L. Ackoff, *Introduction to Operations Research*, New York, John Wiley, 1968. G. Chadwick in *A Systems View of Planning*, New York, Pergamon, 1971, defines a model as "the representation of a system by means of another system." There are very many classifications of models and they vary according to the viewpoint of the person presenting them. Thus, Colin Lee in *Models in Planning*, New York, Pergamon, 1973, limits himself to distinguishing "physical" from "abstract" models. With respect to the latter, he then considers those that are quantitative, that is, expressed in mathematical equations, which, according to him, are "much more important in the context of the aims of planning." Francis Ferguson in *Architecture, Cities and the Systems Approach* New York, Braziller, 1975, divides models into "descriptive" and "predictive" ones. He then considers the latter in reference to simulation processes and maintains that these models, and in relation to such functioning, are especially useful in planning processes.

20. Beer's Chilean work during the Allende years dealt with this issue; see *Platform for Change*, John Wiley: New York, 1974.

21. *Readings in Urban Sociology*, R. E. Pahl, ed., London, 1968.

22. Such a viewpoint also emerges from an article by R. Rossanda, N. Cini and L. Berlinguer in *Manifesto* (a monthly), no. 2, 1970.

23. The study of preindustrial society shows that the range of activities in which a person was then involved was much larger and more flexible and much more dependent on personal attitudes and ability even in very stratified societies. In the modern world diplomas and certificates make flexibility nearly zero although there are movements possible, horizontal as well as vertical, between and among these inflexible posts. Scholars who have studied primitive peoples have brought to light societies based on the near total absence of the division of labor. In this regard see: Colin M. Turnbull, *Wayward Servants* Garden City, N.Y., The Natural History Press, 1965.

24. To better understand how this is really possible, we can for a moment distinguish two principal dimensions on which specialization is built: 1) special knowledge, including the special language on which every specialization rests; 2) the special operations that those possessing that knowledge can presumably perform and are alone capable of performing. As for the first, to eliminate specialization does not mean to destroy everything important that one has learned to understand or build so far, but to accept or to act in a way that such knowledge is shared and not reserved for a few on the false ground that few are able to possess it. Probably before the epoch of mass literacy, the concepts necessary for learning how to read and write were described and made to appear as the inborn property of a few; reserving them for the few served only to produce and maintain privileged classes. As for being able to perform special operations, if we would examine many current professions, protected by the exclusive professional societies, we would conclude that many operations could be performed with equally good results by nonspecialists (including some not especially difficult surgical operations).

25. Lefebvre, *La révolution urbaine*, p. 25.

26. In other words, the fact of reaching a condition of having human beings who gradually succeed in defining themselves and having a self-awareness, and therefore an awareness also of others, as total persons means to create an input into the daily life of a factory, a hospital or a bank. That leads to a progressive and increasing transformational impact on those other institutions.

27. Lefebvre, *La révolution urbaine*, p. 197.

28. *Ibid.*, p. 237.

CHAPTER 5

1. Due to its artistic and historical past, Faenza has accumulated an urban and architectural inheritance of notable importance. Before the Second World War the ancient part within the walls made up the entire urban fabric. There were only a few scattered buildings outside the walls and these were mostly related to agriculture. From the postwar period on people increasingly left the historical center and old houses, too expensive to restore, were abandoned. This and the decreased agricultural activities led to the formation of a relatively large peripheral ring where today the inhabitants are more numerous than those living in the ancient part, in the historical center. Since the postwar period Faenza has not had a notable increase in the total population. The exodus of several thousand people from Faenza, especially the young, has been balanced by the arrival

of people from the small towns in the surrounding foothills and mountains. These areas, like others similar to them throughout Italy, have undergone a remarkable depopulation. Thus, even though Faenza has experienced an exodus of some of its population, people have been attracted to it, although to a much lesser extent than to larger cities in the same region, such as Ravenna or Bologna.

2. Obviously, there are differences. One is that in the historical centers in Italy one never finds the total elimination of residences that often occurs in the many downtown areas of finance, industry, and commerce in America. This is due to the general preference of the European bourgeoisie for urban locations whereas in America their preference was from early colonial times more for rural or small town locations and life styles.

3. In this regard see the issue of *Radical America* dedicated to "The Fiscal Crisis of the City," January 1977.

4. In Italy there is a law that obliges municipalities to have plans signed by at least one professional person from the categories authorized to make them, i.e., architects and engineers. In recent times many municipalities, even the smaller ones, have an urban planning section, but they tend to entrust plans to "independent" professionals (referring to formal occupational independence and not political party independence). These latter usually work alongside an "urban planning commission" nominated by the city council and formed in various ways depending on local politics. As I pointed out earlier, given the economic importance of plans and the speculative game over the different areas, every party generally wants direct "control" over the plans or over the choices concerning the use of the areas. Thus, the parties want professionals who are in their confidence to do the planning. This has led to the absurd formation of planning teams made up of as many members as there are important parties in the municipality.

5. Especially from 1969 to the present, the so-called "battle" for housing in Italy has been one of the dominant themes both of worker agitation and of the writings of urban planners. There are innumerable labor union documents and books on the subject, even if not terribly original.

6. It is important for North American readers to keep in mind that in Italy in local elections as well as national the candidates are party representatives, generally the same parties that are present in national elections. No candidate can run alone. All of the lists of party candidates who are running must collect a minimum number of signatures from supporters and that must be legalized by a notary. That gives parties extraordinary advantages over far less well organized nonparty groups of citizens.

7. The group was formed in this manner: three were designated by the three parties then comprising the local government coalition and one by the major opposition party. Thus the group was made up of a Christian Democrat, a Republican and a Socialist, the three governing coalition parties then, and a Communist, then the opposition party.

8. The selection of the sample in Faenza was done in the following manner. The number of persons desired for the total sample was established (800) and allocated proportionately between those living in the geographical area containing the historical center and the urbanized area adjacent to it as one part of the sample and those living in the rest of the rural county (the commune). In the analysis a further division was made between those living within the historical center and those living in the urbanized periphery and suburbs. The total urbanized area sample was taken from up-to-date electoral lists of the municipality by drawing the name of every Nth person. The second operation was the drawing of the rural sample in two stages. Small areas were sampled and then every

Nth-plus-X person was drawn from the same comprehensive electoral lists once the household addresses were known in this more sparsely populated area. These lists were of persons 18 years of age or over.

9. I am referring here to the examples of Bologna, Pesaro and Ferrara that provide the most famous plans of historical centers in this period. For the first a small social study was made after the plan was made. The social study of the second was made by taking into account only the heads of families and was a traditional investigation of the problem of housing. The third had no social research.

10. See in this regard in the international education journal *Convergence* (vol. VIII, no. 2, 1975) the part entitled "Special Feature: Participatory Research," and especially the article by J. Ohliger and J. Niemi, containing an extended bibliography, pp. 82 ff.

11. The neighborhood district councils or neighborhood councils, literally councils of the quarters, were introduced at the beginning of the sixties in cities that were already controlled by the left, especially by the Communist Party. After a short time they were also introduced in such large cities as Milan, but were then controlled by the Christian Democratic Party. The stated intentions were to "decentralize" power toward the grass roots. For several years the Communists, in those cities in which they were an opposition party, made the realization of the district councils an object of battle often amounting, however, to a purely demagogic exercise. In fact, many on the non-communist left, meaning here the so-called extra-parliamentary groups to the left and also critics of the Communist Party were harsh critics from the start of the forms and results of this version of decentralization that was not, in fact, really one of power decentralization. See, for example, the critical discussion by G. Dalla Pergola, *La Conflittualità Urbana*, Milan, Feltrinelli, 1972.

12. In *Webster's New World Dictionary* (1957 edition) one finds the following definition of the term "Metonymy": "the use of the name of one thing for that of another associated with or suggested by it—(e.g., 'the White House has decided' for 'the President has decided'). "

13. As I have already indicated, the historical center represents that part of the urban area limited by the perimeter of the ancient walls. The periphery or suburbia is that part of the urban area that was formed more recently. The agricultural area makes up those settlements that are scattered but bounded by the administrative confines of the municipality. The boundaries of municipalities, communes, extend in Italy beyond the urban center or urbanized area and resemble American counties in that sense.

14. One can only speculate that the change of those holding office in the local government, having taken place a few months before the investigation and after the administrative elections, helped to create this situation in Faenza. Due to a lack of previous comparative data, one cannot say if it is really a new condition in Faenza nor, assuming it is, can one hypothesize about the stability of such a change. It does, though, provide evidence for one of two possibilities: either the usual belief that Italians are pessimistic or cynical with respect to all forms of government does not correspond to the truth or, granted that this was once generally true, it is not true that this situation is unchangeable. Our feeling is that the latter is the case and that Faenza is not a rare, deviant case.

15. *Old Age*, Harmondsworth, Penguin, p. 16.

16. In order to clarify what I mean, I will give the specific list of initiatives that were brought to the attention of those interviewed:

I) To provide Faenza with open areas and places equipped for gatherings of citizens of various ages (the elderly and young people) and of various categories (housewives, students, etc.) so as to permit them to have social exchanges.

II) To close either all or part of the center of the city to car traffic, only permitting the coming and going of the cars of inhabitants and in certain hours the movement of goods according to the needs of the shops in the business section.

III) To foresee or facilitate the creation of flower or vegetable gardens that could be used in common by a number of families living in the surrounding areas.

IV) To create a number of activities for the elderly in the historical center, giving to those elderly who wish it the chance to do socially useful although not heavy work.

V) To integrate the schools with the life of the community by using them for cultural, educational and social initiatives and activities for adults at the district and neighborhood level.

VI) To arrange for the building of public housing in the historical center so as to guarantee decent living quarters also for those who cannot pay high rent (workers, young couples, etc.).

VII) To guarantee access to public housing for the elderly that will provide them not only decent housing but also to be in contact with the everyday life of the community.

17. One of the matters that began to be discussed by citizens who ordinarily felt this to be a matter for private development, the Church, or at most local officials and not the people at large was specific to Faenza. (All cities in all countries, however, have similar matters of equal, potential interest to citizens.) In Faenza this matter was the use of Church properties in the historical center. Before the unification of Italy in 1861, Faenza was under papal government, as was a large part of the area called Romagna and as was Bologna. Probably due partly to geographical and partly to political reasons, Faenza had become a focal point for many Catholic religious communities. Especially in the historical center, this resulted in a notable part of the property belonging to these religious orders, and this is still the case today. Some of these groups suffer from the so-called "crisis of vocation" besetting the entire Catholic Church and, like all religious structures in Italy, they are also losing the dominant role that until now they have had in political control of the society. Large areas of land connected to huge buildings, in some cases no longer used, could bring extremely large economic profits if used for high-density building rather than public areas. These alternative uses became a matter of active concern to the citizens as our participatory process unfolded.

CHAPTER 6

1. Such price or cost self-reduction took various forms of consumer direct actions. A wide array developed far beyond the classic rent strike. Advocacy of direct action in the American urban crisis is found, for example, in Frances Fox Piven and Richard A. Cloward in "The Urban Crisis as an Arena for Class Mobilization," *Radical America*, vol. 11, no. 1, pp. 9-17.

2. For such a position see Norton E. Long, "Have Cities a Future?", *Public Administration Review*, November-December 1973, pp. 543-552.

3. The most interesting exposition can be found in Kenneth Clark and Jeannette Hopkins, *A Relevant War Against Poverty*, New York, Harper & Row, 1970.

4. Nathan Glazer in a reasoned defense of community control makes the point correctly that decentralization is not the same as community control. The latter means the former, but not vice versa. See his "For White and Black, Com-

munity Control is the Issue," *Metropolitan Communities*, ed. by Joseph Bensman and Arthur J. Vidich, New York: A New York Times Book, New View-
points, 1975, pp. 197-212.

5. See the statement of the public choice approach in Robert L. Bish and
Vincent Ostrom, *Understanding Urban Government*, Washington, D.C., American Enterprise Institute for Public Policy Research, 1973, pp. 17-34.

6. Christopher Alexander stresses how mental habits, the way the mind
works, have trapped designers into conceiving of cities as trees rather than semi-
lattices. I suggest that this is operating also for his own mind, and we can expect
it to happen with the minds of most urban planners, in regard to the central
point of our thesis. Alexander cannot seem to give up his own self-identifying
coordinates as a professional planner. One is the necessity of seeing "from the
designer's point of view, the physically unchanging part of this system [as] of
special interest." This is the network of "the unchanging receptacles," the "fixed
part of the system," and the basic units of the city: "I define this fixed part as a
unit of the city." At the same time, he recognizes more than most that the physical and its meanings change as people change, as history unfolds. Yet he cannot
give up his mind-set as he includes people in the *physical* aspect of the city: ". . .
as designers, we are concerned with the physical living city and its physical back-
bone, we most naturally restrict ourselves to considering sets which are collec-
tions of material elements such as people, blades of grass, cars, bricks. . . ."

Nor, despite a host of insights leading him to properly reject the before, by
and after Corbusier traditions of segregated zoning and functional sectors, can
Alexander seem to rid himself of his second basic coordinate, the special task for
designers, for professionals, for urban planners to sketch or map "the right" city
for the present and future, a city of overlapping places and activities. He cannot,
or could not in 1965, conceive of a planning process that involves the masses of
people as participants in an opened up planning process, in a process that permits the appropriate sociophysical forms of the city to emerge from socio-phys-
ical participation in the life spaces of the city and in a popular self-reflection
with merely more skilled architects, artists, and designers of the past and potential futures. And to have "the right overlap," as Alexander puts it, of spatially
located human activities in a city will forever escape the mind's eye of the designer who sees it as his God-given task to think it or draw it because of his special expertise and who cannot shed his physical and social, physical and aesthetic,
physical and nonphysical dichotomies. He must somehow become a teacher for
masses of people who, he can hope, *become with him* persons of learning and of
civility and of respected teaching capabilities, for teaching themselves and others
in the future about good living and, hence, about good cities. See "A City is Not a
Tree," *Architectural Forum*, Parts I and II, 122 (April/May 1965), pp. 58-62.

7. For descriptions of such community unions and what they have been doing, see *Mountain Life & Work*, a monthly published by the Council of the
Southern Mountains, Inc., Drawer N, Clintwood, Va. 24228.

8. That is documented by Morton and Lucia White, *The Intellectual Versus
the City*, Cambridge, Mass., Harvard University Press and MIT Press, 1962.

9. For a persuasive picture of what is likely to happen in the absence of dramatic innovations, see Robert Heilbroner, *An Inquiry into the Human Prospect*,
New York, W. W. Norton, 1975.

10. A marvelous example of at least semimaieutic planning is the case of A.
Fathy, an Egyptian architect-planner. Although, as in Faenza, his experience finally had an outcome that can only be described as a failure, his report of the
unsuccessful experience is a must for the future maieutic planner of any country.
See his *Architecture for the Poor*, Chicago, University of Chicago Press, 1973. Or

see the documentary film made for CBC television on Fathy's effort by Kathy Smalley of the program series *Man Alive* (CBC, Toronto, Canada).

11. The relevant and sad experiences reported by Karl Hess in the abortive efforts of ordinary citizens, not professionals, to renew, revitalize and redevelop in different directions than usual a large neighborhood of about 40,000 people in Washington, D.C., must not be ignored. There are other such experiences and they would seem to contradict any optimism about the possibilities raised in this book about different and innovative directions in the United States in terms of opening such institutions as urban planning and starting to create urban self-management.

We have no simple recipes nor do we think that such tasks are easy. What appears to be almost inner character-structure needs of dependency, desires for services and preferences for representation rather than participation will, we think, prove to be transmutable attitudes, just as the deep wariness and caution of some people will give way, once a new set of experiences begin to happen and to feed back in a way reinforcing their developing self-confidence. Exactly what kind or amount of "critical mass" may be necessary before such processes start we cannot say. But we suspect that they must involve changes in atmospheres affecting more than single neighborhoods, perhaps throughout a whole state or set of states or a profession or a set of professions or a complex of people in various places. Certainly urban planners and students of urban planning are still protected or partly protected by the college and university atmospheres that the lonely practitioner in the front line will need to turn to periodically for support and energies until the process becomes self-sustaining. See Karl Hess, "Flight from Freedom: Memories of a Noble Experiment," *Quest Magazine*, September/October, pp. 39-46.

12. The progressive change requires social suicide but provides time for metamorphosis of planning roles and, indeed, of self-identities. Thus, planners can join with others to create a communal organization, which resolves the contradiction, underlined by Shimon S. Gottschalk, of planners being unable to plan communal organizations because the latter "must generate themselves." Because communal organizations are people, and the planner has the (human) right to act as a person rather than a planner in narrow professional terms, he would be joining with others in a joint venture—escaping the dilemma noted by Gottschalk of his having to serve the status quo or do nothing. See his *Communities and Alternatives: An Exploration of the Limits of Planning*, New York, Wiley & Sons, 1975, p. 126. This is our effort to escape from the eternal radical conundrum explicated by Frances Fox Piven and Richard Cloward in *Poor People's Movements*, New York, Pantheon, 1977.

ABOUT THE AUTHOR

Simona Ganassi Agger is Professor of Architecture at the Istituto Universitario Architettura Venezia, and Professor of Sociology at Eastern Kentucky University. She has extensive international experience as an urban researcher and planner, and for five years served as president of the Cooperative Center for Planning and Projects in Venice, Italy. She is now affiliated with the International Centre for Human Community in Venice. The present study is a revised version of her book *Autogestione Urbana: L'Urbanistica per una Nuova Società*, published by Dedalo in 1977.